FRACTIONS, DECIMALS, & PERCENTS

GRE Math Preparation Guide

This book provides an in-depth look at the array of GRE questions that test knowledge of Fractions, Decimals, and Percents. Learn to see the connections among these part–whole relationships and practice implementing strategic shortcuts.

Fractions, Decimals, & Percents GRE Strategy Guide, Second Edition

10-digit International Standard Book Number: 1-935707-48-5
13-digit International Standard Book Number: 978-1-935707-48-6
eISBN: 978-0-984178-07-0

8 GUIDE INSTRUCTIONAL SERIES

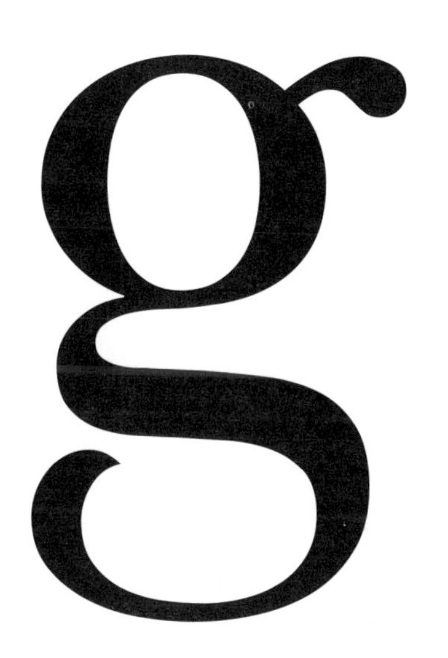

Math GRE Preparation Guides

Algebra
(ISBN: 978-1-935707-47-9)

Fractions, Decimals, & Percents
(ISBN: 978-1-935707-48-6)

Geometry
(ISBN: 978-1-935707-49-3)

Number Properties
(ISBN: 978-1-935707-50-9)

Word Problems
(ISBN: 978-1-935707-54-7)

Quantitative Comparisons & Data Interpretation
(ISBN: 978-1-935707-51-6)

Verbal GRE Preparation Guides

Reading Comprehension & Essays
(ISBN: 978-1-935707-52-3)

Text Completion & Sentence Equivalence
(ISBN: 978-1-935707-53-0)

March 15th, 2011

Dear Student,

Thank you for picking up one of the Manhattan GRE Strategy Guides—we hope that it ends up being just what you need to prepare for the new GRE.

As with most accomplishments, there were many people involved in the book that you're holding. First and foremost is Zeke Vanderhoek, the founder of MG Prep. Zeke was a lone tutor in New York when he started the Company in 2000. Now, eleven years later, the Company has Instructors and offices nationwide and contributes to the studies and successes of thousands of students each year.

Our Manhattan GRE Strategy Guides are based on the continuing experiences of our Instructors and our students. On the Company side, we are indebted to many of our Instructors, including but not limited to Roman Altshuler, Chris Berman, Faruk Bursal, Jen Dziura, Dmitry Farber, Stacey Koprince, David Mahler, Seb Moosapoor, Stephanie Moyerman, Chris Ryan, Michael Schwartz, Tate Shafer, Emily Sledge, Tommy Wallach, and Ryan Wessel, all of whom either wrote or edited the books to their present form. Dan McNaney and Cathy Huang provided their formatting expertise to make the books as user-friendly as possible. Last, many people, too numerous to list here but no less appreciated, assisted in the development of the online resources that accompany this guide.

At Manhattan GRE, we continually aspire to provide the best Instructors and resources possible. We hope that you'll find our dedication manifest in this book. If you have any comments or questions, please e-mail me at dan@manhattangre.com. I'll be sure that your comments reach our curriculum team—and I'll read them too.

Best of luck in preparing for the GRE!

Sincerely,

Dan

Dan Gonzalez
Managing Director
Manhattan GRE

www.manhattangre.com 138 West 25th St., 7th Floor NY, NY 10001 Tel: 212-721-7400 Fax: 646-514-7425

HOW TO ACCESS YOUR ONLINE STUDENT CENTER

If you...

are a registered Manhattan GRE student

and have received this book as part of your course materials, you have AUTOMATIC access to ALL of our online resources. To access these resources, follow the instructions in the Welcome Guide provided to you at the start of your program. Do NOT follow the instructions below.

purchased this book from the Manhattan GRE Online store or at one of our Centers

1. Go to: http://www.manhattangre.com/studentcenter.cfm

2. Log in using the username and password used when your account was set up.
Your one year of online access begins on the day that you purchase the book from the Manhattan GRE online store or at one of our centers.

purchased this book at a retail location

1. Go to: http://www.manhattangre.com/access.cfm

2. Log in or create an account.

3. Follow the instructions on the screen.

Your one year of online access begins on the day that you register your book at the above URL.

You only need to register your product ONCE at the above URL. To use your online resources any time AFTER you have completed the registration process, login to the following URL:
http://www.manhattangre.com/studentcenter.cfm

Please note that online access is non-transferable. This means that only NEW and UNREGISTERED copies of the book will grant you online access. Previously used books will not provide any online resources.

purchased an e-book version of this book

1. Create an account with Manhattan GRE at the website:
   https://www.manhattangre.com/createaccount.cfm

2. Email a copy of your purchase receipt to books@manhattangre.com to activate your resources. Please be sure to use the same email address to create an account that you used to purchase the e-book.

For any technical issues, email books@manhattangre.com or call 646-254-6479.

TABLE OF CONTENTS

g

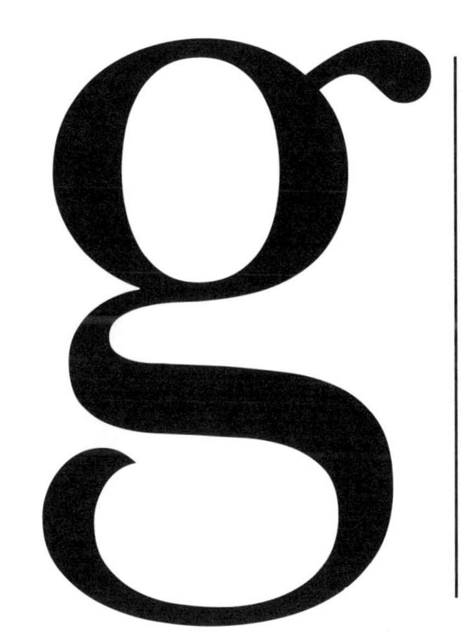

Chapter 1
of
FRACTIONS, DECIMALS, & PERCENTS

INTRODUCTION & THE REVISED GRE

In This Chapter . . .

- Introduction, and How to Use Manhattan GRE's Strategy Guides
- The Revised GRE
- Question Formats in Detail

Introduction, and How to Use Manhattan GRE's Strategy Guides

We know that you're looking to succeed on the GRE so that you can go to graduate school and do the things you want to do in life.

We also know that you might not have done math since high school, and that you may never have learned words like "adumbrate" or "sangfroid." We know that it's going to take hard work on your part to get a top GRE score, and that's why we've put together the only set of books that will take you from the basics all the way up to the material you need to master for a near-perfect score, or whatever your score goal may be. You've taken the first step. Now it's time to get to work!

How to Use These Materials

Manhattan GRE's materials are comprehensive. But keep in mind that, depending on your score goal, it may not be necessary to "get" absolutely everything. Grad schools only see your overall Quantitative, Verbal, and Writing scores—they don't see exactly which strengths and weaknesses went into creating those scores.

You may be enrolled in one of our courses, in which case you already have a syllabus telling you in what order you should approach the books. But if you bought this book online or at a bookstore, feel free to approach the books—and even the chapters within the books—in whatever order works best for you. *For the most part, the books, and the chapters within them, are independent; you don't have to master one section before moving on to the next.* So if you're having a hard time with something in particular, you can make a note to come back to it later and move on to another section. Similarly, it may not be necessary to solve every single practice problem for every section. As you go through the material, continually assess whether you understand and can apply the principles in each individual section and chapter. The best way to do this is to solve the Check Your Skills and Practice Problems throughout. If you're confident you have a concept or method down, feel free to move on. If you struggle with something, make note of it for further review. Stay active in your learning and oriented toward the test—it's easy to read something and think you understand it, only to have trouble applying it in the 1–2 minutes you have to solve a problem.

Study Skills

As you're studying for the GRE, try to integrate your learning into your everyday life. For example, vocabulary is a big part of the GRE, as well as something you just can't "cram" for—you're going to want to do at least a little bit of vocab every day. So, try to learn and internalize a little bit at a time, switching up topics often to help keep things interesting.

Keep in mind that, while many of your study materials are on paper (including ETS's most recent source of official GRE questions, *The Official Guide to the GRE revised General Test*), your exam will be administered on a computer. Because this is a computer-based test, you will NOT be able to underline portions of reading passages, write on diagrams of geometry figures, or otherwise physically mark up problems. So get used to this now. Solve the problems in these books on scratch paper. (Each of our books talks specifically about what to write down for different problem types).

Again, as you study stay focused on the test-day experience. As you progress, work on timed drills and sets of questions. Eventually, you should be taking full practice tests (available at www.manhattangre.com) under actual timed conditions.

The Revised GRE

As of August 1, 2011, the Quantitative and Verbal sections of the GRE will undergo a number of changes. The actual body of knowledge being tested won't change, but the *way* it is tested will. Here's a brief summary of what to expect, followed by a more comprehensive assessment of the new exam.

Overall, the general format of the test will change. The length of the test will increase from about 3.5 hours to about 4 hours. There will be two scored math sections and two scored verbal sections rather than one of each, and a new score scale of 130–170 will be used in place of the old 200–800 scale. More on this later.

The Verbal section of the GRE will change dramatically. The Analogies and Antonym questions will disappear. The Sentence Completions and Reading Comprehension will remain, to be expanded and remixed in a few new ways. Vocabulary will still be important, but only in the context of complete sentences. That is, you'll no longer have to worry about vocabulary words standing alone. So for those who dislike learning vocabulary words, the changes will provide partial relief. For those who were looking forward to getting lots of points just for memorizing words, the Manhattan GRE verbal strategy guides will prepare you for the shift.

The Quant section of the GRE prior to August 1, 2011 is composed of multiple choice problems, Quantitative Comparisons, and Data Interpretation questions (which are really a subset of multiple choice problems). The revised test will contain two new problem formats in addition to the current problem formats. However, the type of math, and the difficulty of the math, will remain unchanged.

Additionally, a small four-function calculator with a square root button will appear on-screen. Many test takers will rejoice at the advent of this calculator! It is true that the GRE calculator will reduce emphasis on computation—but look out for problems, such as percents questions with tricky wording, that are likely to foil those who rely on the calculator too much. *In short, the calculator may make your life a bit easier from time to time, but you will never need the calculator to solve a problem.*

Finally, don't worry about whether these new problem types are "harder" or "easier." You are being judged against other test takers, all of whom are in the same boat. So if the new formats are harder, they are harder for other test takers as well.

Exam Structure

The revised test has six sections. You will get a ten-minute break between the third and fourth sections and a one-minute break between the others. The Analytical Writing section is always first. The other five sections can be seen in any order and will include:

- Two Verbal Reasoning sections (approximately 20 questions each in exactly 30 minutes per section)
- Two Quantitative Reasoning sections (approximately 20 questions each in exactly 35 minutes per section)
- Either an "unscored" section or a "research" section

An unscored section will look just like a third Verbal or Quantitative Reasoning section, and you will not be told which of them doesn't count. If you get a research section, it will be identified as such.

	Section Type	# Questions	Time	Scored?
	Analytical Writing	2 essays	30 minutes each	Yes
	Verbal #1	Approx. 20	30 minutes	Yes
	Quantitative #1	Approx. 20	35 minutes	Yes
10 min break →	Verbal #2	Approx. 20	30 minutes	Yes
	Quantitative #2	Approx. 20	35 minutes	Yes
one or the other, but not both	Unscored Section (verbal or quant)	Approx. 20	30 or 35 min	No
	Research Section	Varies	Varies	No

Later in the chapter, we'll look at all the question formats in detail.

Using the Calculator

The addition of a small, four-function calculator with a square root button means that those taking the revised test can forget re-memorizing their times tables or square roots. However, the calculator is not a cure-all; in many problems, the difficulty is in figuring out what numbers to put into the calculator in the first place. In some cases, using a calculator will actually be less helpful than doing the problem some other way. Let's look at an example:

If x is the remainder when (11)(7) is divided by 4 and y is the remainder when (14)(6) is divided by 13, what is the value of $x + y$?

Solution: This problem is designed so that the calculator won't tell the whole story. Certainly the calculator will tell us that $11 \times 7 = 77$. When you divide 77 by 4, however, the calculator yields an answer of 19.25. The remainder is *not* 0.25 (a remainder is always a whole number).

You might just go back to your pencil and paper, and find the largest multiple of 4 that is less than 77. Since 4 DOES go into 76, we can conclude that 4 would leave a remainder of 1 when dividing into 77. (Notice that we don't even need to know how many times 4 goes into 76, just that it goes in. One way to mentally "jump" to 76 is to say, *4 goes into 40, so it goes into 80 … that's a bit too big, so take away 4 to get 76.*) You could also multiply the leftover 0.25 times 4 (the divisor) to find the remainder of 1.

However, it is also possible to use the calculator to find a remainder. Divide 77 by 4 to get 19.25. Thus, 4 goes into 77 nineteen times, with a remainder left over. Now use your calculator to multiply 19 (JUST 19, not 19.25) by 4. You will get 76. The remainder is $77 - 76 = 1$. Therefore, $x = 1$.

Use the same technique to find y. Multiply 14×6 to get 84. Divide 84 by 13 to get 6.46... Ignore everything after the decimal, and just multiply 6 by 13 to get 78. The remainder is therefore $84 - 78 = 6$. Therefore, $y = 6$.

Since we are looking for $x + y$ and $1 + 6 = 7$, the answer is 7.

You can see that blind faith in the calculator can be dangerous. Use it responsibly! And this leads us to...

Practice Using the Calculator!

On the new GRE, the on-screen calculator will slow you down or lead to incorrect answers if you're not careful! If you plan to use the thing on test day (which you should), you'll want to pactice first.

We have created an online practice calculator for your use. To access this calculator, go to www.manhattangre.com and sign in to the student center using the instructions on the "How to Access Your Online Student Center" page found at the front of this book.

In addition to the calculator, you will see instructions for how to use the calculator. Be sure to read these instructions and work through the associated exercises. Throughout our math books, you will see the symbol. This symbol means "use the calculator here!" As much as possible, have the online practice calculator up and running during your review of our math books. You'll have the chance to use the on-screen calculator when you take our practice exams as well.

Navigating the Questions in a Section

Another change for test takers on the new GRE is the ability to move freely around the questions in a section... you can go forward and backward one-by-one and can even jump directly to any question from the "review list." The review list provides a snapshot of which questions you have answered, which ones you have tagged for "mark and review," and which are incomplete, either because you didn't select enough answers or because you selected too many (that is, if a number of choices is specified by the question). You should double-check the review list for completion if you finish the section early. Using the review list feature will take some practice as well, which is why we've built it into our online practice exams. Here's some introductory advice.

The majority of test takers will be pressed for time. Thus, for most of you, it won't be feasible to "go back to" multiple problems at the end of the section. Generally, if you can't get a question the first time, you won't be able to get it the second time around either. With this in mind, here's how we recommend using the new review list feature.

1. Do the questions in order as they appear.

2. When you encounter a difficult question, do you best to eliminate answer choices you know are wrong.

3. If you're not sure of an answer, take an educated guess from the choices remaining. Do <u>NOT</u> skip it and hope to return to it later.

4. Using the "mark" button at the top of the screen, mark up to three questions per section that you think you might be able to solve with more time. Mark a question only after you have taken an educated guess.

5. If you have time at the end of the section, click on the review list, identify any questions you've marked and return to them. If you do not have any time remaining, you will have already taken good guesses at the tough ones.

What you want to avoid is "surfing"—clicking forward and backward through the questions searching for the easy ones. This will eat up valuable time. Of course, you'll want to move through the tough ones quickly if you can't get them, but try to avoid skipping stuff.

Again, all of this will take practice. Use our practice exams to fine-tune your approach.

Scoring

Two things have changed about the scoring of the Verbal Reasoning and Quantitative Reasoning sections: (1) how individual questions influence the score and (2) the score scale itself.

For both the Verbal Reasoning and Quantitative Reasoning sections, you will receive a raw score, which is simply how many questions you answered correctly. Your raw score is converted to a scaled score, accounting for the difficulties of the specific questions you actually saw.

The old GRE was question-adaptive, meaning that your answer to each question (right or wrong) determined, at least somewhat, the questions that followed (harder or easier). Because you had to commit to an answer to let the algorithm do its thing, you weren't allowed to skip questions or go back to change answers. On the revised GRE, the adapting will occur from section to section (e.g., if you do well on the first verbal section, you will get a harder second verbal section) rather than from question to question. The only change test takers will notice is one most will welcome: you can now move freely about the questions in a section, coming back to tough questions later, changing answers after "ah-ha!" moments, and generally managing your time more flexibly.

The scores for the revised GRE Quantitative Reasoning and Verbal Reasoning will be reported on a 130 to 170 scale in 1-point increments, whereas the old score reporting was on a 200 to 800 scale in 10-point increments. You will receive one 130–170 score for verbal and a separate 130–170 score for quant. If you are already putting your GRE math skills to work, you may notice that there are now 41 scores possible (170 – 130, then add one before you're done), whereas before there were 61 scores possible ([800 – 200]/10, then add one before you're done). In other words, a 10 point difference on the old score scale actually indicated a smaller performance differential than a 1 point difference on the new scale. However, the GRE folks argue that perception is reality: the difference between 520 and 530 on the old scale could simply seem greater than the difference between 151 and 152 on the new scale. If that's true, then this change will benefit test-takers, who won't be unfairly compared by schools for minor differences in performance. If not true, then the change will be moot.

Important Dates

Registration for the GRE revised General Test opens on March 15, 2011, and the first day of testing with the new format is August 1, 2011.

Perhaps to encourage people to take the revised exam, rather than rushing to take the old exam before the change or waiting "to see what happens" with the new exam long after August 1, 2011, ETS is offering a 50% discount on the test fee for anyone who takes the revised test from August 1 through September 30, 2011. Scores for people who take the revised exam in this discount period will be sent starting in mid- to late-November. This implies that you may have to wait up to 3.5 months to get your score during this rollout period!

By December 2011, ETS expects to resume normal score reporting schedules: score reports will be sent a mere 10-15 days after the test date.

IMPORTANT: If you need GRE scores before mid-November 2011 to meet a school deadline, take the "old" GRE no later than July 31, 2011! Waiting to take the revised test not only would require you to study for a different test, but also would delay your score reporting.

Question Formats in Detail

Essay Questions

The Analytical Writing section consists of two separately timed 30-minute tasks: Analyze an Issue and Analyze an Argument. As you can imagine, the 30-minute time limit implies that you aren't aiming to write an essay that would garner a Pulitzer Prize nomination, but rather to complete the tasks adequately and according to the directions. Each essay is scored separately, but your reported essay score is the average of the two rounded up to the next half-point increment on a 0 to 6 scale.

Issue Task—This essay prompt will present a claim, generally one that is vague enough to be interpreted in various ways and discussed from numerous perspectives. Your job as a test taker is to write a response discussing the extent to which you agree or disagree and support your position. Don't sit on the fence—pick a side!

For some examples of Issue Task prompts, visit the GRE website here:

http://www.ets.org/gre/revised_general/prepare/analytical_writing/issue/pool

Argument Task—This essay prompt will be an argument comprised of both a claim(s) and evidence. Your job is to dispassionately discuss the argument's structural flaws and merits (well, mostly the flaws). Don't agree or disagree with the argument—evaluate its logic.

For some examples of Argument Task prompts, visit the GRE website here:

http://www.ets.org/gre/revised_general/prepare/analytical_writing/argument/pool

Verbal: Reading Comprehension Questions

Standard 5-choice multiple choice reading comprehension questions will continue to appear on the new exam. You are likely familiar with how these work. Let's take a look at two *new* reading comprehension formats that will appear on the new test.

Select One or More Answer Choices and Select-in-Passage

For the question type, "Select One or More Answer Choices," you are given three statements about a passage and asked to "select all that apply." Either one, two, or all three can be correct (there is no "none of the above" option). There is no partial credit; you must select all the correct choices and none of the incorrect choices.

> *Strategy Tip: On "Select One or More Answer Choices," don't let your brain be tricked into telling you "Well, if two of them have been right so far, the other one must be wrong," or any other arbitrary idea about how many of the choices "should" be correct. Make sure to consider each choice independently! You cannot use "Process of Elimination" the same way as you do on "normal" multiple-choice questions.*

For the question type "Select-in-Passage," you are given an assignment such as "Select the sentence in the passage that explains why the experiment's results were discovered to be invalid." Clicking anywhere on the sentence in the passage will highlight it. (As with any GRE question, you will have to click "Confirm" to submit your answer, so don't worry about accidentally selecting the wrong sentence due to a slip of the mouse.)

> *Strategy Tip: On "Select-in-Passage," if the passage is short, consider numbering each sentence (that is, writing 1 2 3 4 on your paper) and crossing off each choice as you determine that it isn't the answer. If the passage is long, you might write a number for each paragraph (I, II, III), and tick off each number as you determine that the correct sentence is not located in that paragraph.*

Now let's give these new question types a try!

The sample questions below are based on this passage:

> Physicist Robert Oppenheimer, director of the fateful Manhattan Project, said "It is a profound and necessary truth that the deep things in science are not found because they are useful; they are found because it was possible to find them." In a later address at MIT, Oppenheimer presented the thesis that scientists could be held only very nominally responsible for the consequences of their research and discovery. Oppenheimer asserted that ethics, philosophy, and politics have very little to do with the day-to-day work of the scientist, and that scientists could not rationally be expected to predict all the effects of their work. Yet, in a talk in 1945 to the Association of Los Alamos Scientists, Oppenheimer offered some reasons why the Manhattan project scientists built the atomic bomb; the justifications included "fear that Nazi Germany would build it first" and "hope that it would shorten the war."

For question #1, consider each of the three choices separately and select all that apply.

1. The passage implies that Robert Oppenheimer would most likely have agreed with which of the following views:

[A] Some scientists take military goals into account in their work
[B] Deep things in science are not useful
[C] The everyday work of a scientist is only minimally involved with ethics

2. Select the sentence in which the writer implies that Oppenheimer has not been consistent in his view that scientists have little consideration for the effects of their work.

[Here, you would highlight the appropriate sentence with your mouse. Note that there are only four options.]

Solutions:

1. {A, C} Oppenheimer says in the last sentence that one of the reasons the bomb was built was scientists' "hope that it would shorten the war." Thus, Oppenheimer would likely agree with the view that "Some scientists take military goals into account in their work." B is a trap answer using familiar language from the passage. Oppenheimer says that scientific discoveries' possible usefulness is not why scientists make discoveries; he does not say that the discoveries aren't useful. Oppenheimer specifically says that ethics has "very little to do with the day-to-day work of the scientist," which is a good match for "only minimally involved with ethics."

   Strategy Tip: On "Select One or More Answer Choices," write ABC on your paper and mark each choice with a check, an X, or a symbol such as ~ if you're not sure. This should keep you from crossing out all three choices and having to go back (at least one of the choices must be correct). For example, let's say that on a different question you had marked

   A. X
   B. X
   C. ~

   The one you weren't sure about, (C), is likely to be correct, since there must be at least one correct answer.

2. The correct sentence is: **Yet, in a talk in 1945 to the Association of Los Alamos Scientists, Oppenheimer offered some reasons why the Manhattan project scientists built the atomic bomb; the justifications included "fear that Nazi Germany would build it first" and "hope that it would shorten the war."** The word "yet" is a good clue that this sentence is about to express a view contrary to the views expressed in the rest of the passage.

Verbal: Text Completion Questions

Text Completions are the new, souped-up Sentence Completions. They can consist of 1–5 sentences with 1–3 blanks. When Text Completions have two or three blanks, you will select words for those blanks independently. There is no partial credit; you must make every selection correctly.

Because this makes things a bit harder, the GRE has kindly reduced the number of possible choices per blank from five to three. Here is an old two-blank Sentence Completion, as it would appear on the old GRE:

Old Format:

Leaders are not always expected to ____________ the same rules as are those they lead; leaders are often looked up to for a surety and presumption that would be viewed as ____________ in most others.

A. obey ... avarice

B. proscribe ... insalubriousness

C. decree ... anachronism

D. conform to ... hubris

E. follow ... eminence

And here's how this same sentence would appear on the new exam.

New Format:

> Leaders are not always expected to (i) ____________the same rules as are those they lead; leaders are often looked up to for a surety and presumption that would be viewed as (ii) ____________ in most others.

Blank (i)
decree
proscribe
conform to

Blank (ii)
hubris
avarice
anachronism

On the new GRE, you will select your two choices by actually clicking and highlighting the words you want.

Solution:

In the first blank, we need a word similar to "follow." In the second blank, we need a word similar to "arrogant." Only choice D works in the old format; in the new format, the answer is still "conform to" and "hubris," but you'll make the two choices separately.

Note that in the "Old Format" question, if you knew that you needed a word in the second blank that meant something like "arrogant," and you knew that "hubris" was the only word in the second column with the correct meaning, you could pick correct answer choice D without even considering the first word in each pair. In the new format, this strategy is no longer available to us.

Also note that, in the "Old Format" question, "obey," "conform to," and "follow" mean basically the same thing. On the new GRE, this can't happen: since you select each word independently, no two choices can be synonyms (otherwise, there would be two correct answers).

Strategy Tip: As on the old GRE, do NOT look at the answer choices until you've decided for yourself, based on textual clues actually written in the sentence, what kind of word needs to go in each blank. Only then should you look at the choices and eliminate those that are not matches.

Let's try an example with three blanks.

> For Kant, the fact of having a right and having the (i) _______ to enforce it via coercion cannot be separated, and he asserts that this marriage of rights and coercion is compatible with the freedom of everyone. This is not at all peculiar from the standpoint of modern political thought—what good is a right if its violation triggers no enforcement (be it punishment or (ii) __________)? The necessity of coercion is not at all in conflict with the freedom of everyone, because this coercion only comes into play when someone has (iii)____________ someone else.

Blank (ii)
technique
license
prohibition

Blank (ii)
amortization
reward
restitution

Blank (iii)
questioned the hypothesis of
violated the rights of
granted civil liberties to

Solution:

In the first sentence, use the clue "he asserts that this marriage of rights and coercion is compatible with the freedom of everyone" to help fill in the first blank. Kant believes that "coercion" is "married to" rights and is compatible with freedom for all. So we want something in the first blank like "right" or "power." Kant believes that rights are meaningless without enforcement. Only the choice "license" can work (while a "license" can be physical, like a driver's license, "license" can also mean "right").

The second blank is part of the phrase "punishment or __________," which we are told is the "enforcement" resulting from the violation of a right. So the blank should be something, other than punishment, that constitutes enforcement against someone who violates a right. (More simply, it should be something bad!) Only "restitution" works. Restitution is compensating the victim in some way (perhaps monetarily or by returning stolen goods).

In the final sentence, "coercion only comes into play when someone has __________ someone else." Throughout the text, "coercion" means enforcement against someone who has violated the rights of someone else. The meaning is the same here. The answer is "violated the rights of."

The complete and correct answer is this combination:

Blank (i)	Blank (ii)	Blank (iii)
license	restitution	violated the rights of

In theory, there are $3 \times 3 \times 3 = 27$ possible ways to answer a 3-blank Text Completion—and only one of those 27 ways is correct. The guessing odds will go down, but don't be intimidated. Just follow the basic process: come up with your own filler for each blank, and match to the answer choices. If you're confused by this example, don't worry! We'll start from the beginning in our *Text Completion & Sentence Equivalence* strategy guide.

Strategy Tip: As on the old GRE, do NOT "write your own story." The GRE cannot give you a blank without also giving you a clue, physically written down in the passage, telling you what kind of word or phrase MUST go in that blank. Find that clue. You should be able to give textual evidence for each answer choice you select.

Verbal: Sentence Equivalence Questions

In this question type, you are given one sentence with a single blank. There are six answer choices, and you are asked to pick TWO choices that fit the blank and are alike in meaning.

Of the new question types, this one depends the most on vocabulary and also yields the most to strategy.

No partial credit is given on Sentence Equivalence; both correct answers must be selected. When you pick two of six choices, there are 15 possible combinations of choices, and only one is correct. However, this is not nearly as daunting as it sounds.

Think of it this way—if you have six choices, but the two correct ones must be "similar in meaning," then you have, at most, three possible PAIRS of choices. Maybe fewer, since not all choices are guaranteed to have a "partner." If you can match up the "pairs," you can seriously narrow down your options.

Here is a sample set of answer choices:

[A] tractable

[B] taciturn

[C] arbitrary

[D] tantamount

[E] reticent

[F] amenable

We haven't even given you the question here, because we want to point out how much you can do with the choices alone, if you have studied vocabulary sufficiently.

TRACTABLE and AMENABLE are synonyms (tractable, amenable people will do whatever you want them to do). TACITURN and RETICENT are synonyms (both mean "not talkative"). ARBITRARY (based on one's own will) and TANTAMOUT (equivalent) are not similar in meaning and therefore cannot be a pair. Therefore, the ONLY possible answers are {A, F} and {B, E}. We have improved our chances from 1 in 15 to a 50/50 shot without even reading the question!

Of course, in approaching a Sentence Equivalence, we do want to analyze the sentence the same way we would with a Text Completion—read for a textual clue that tells you what type of word MUST go in the blank. Then look for a matching pair.

Strategy Tip: If you're sure that a word in the choices does NOT have a partner, cross it out! For instance, if A and C are partners, and E and F are partners, and you're sure B and D are not each other's partners, cross out B and D completely. They cannot be the answer together, nor can either one be part of the answer.

The sentence for the answer choice above could read,

> Though the dinner guests were quite ______ , the hostess did her best to keep the conversation active and engaging.

Thus, B and E are the best choices. Let's try an example.

While athletes usually expect to achieve their greatest feats in their teens or twenties, opera singers don't reach the ____________ of their vocal powers until middle age.

[A] harmony

[B] zenith

[C] acme

[D] terminus

[E] nadir

[F] cessation

Solution:

Those with strong vocabularies might go straight to the choices to make pairs. ZENITH and ACME are synonyms, meaning "high point, peak." TERMINUS and CESSATION are synonyms, meaning "end." NADIR is a low point and HARMONY is present here as a trap answer reminding us of opera singers. *Cross off A and E, since they do not have partners.* Then, go back to the sentence, knowing that your only options are a pair meaning "peak" and a pair meaning "end."

The answer is {B, C}.

Math: Quantitative Comparison

This format is a holdover from the old exam. Here's a quick example:

Quantity A	Quantity B
x	x^2

(A) Quantity A is greater.
(B) Quantity B is greater.
(C) The two quantities are equal.
(D) The relationship cannot be determined from the information given.

Solution: If $x = 0$, the quantities are equal. If $x = 2$, quantity B is greater. Thus, we don't have enough information.

The answer is D.

Let's look at the new math question formats.

Math: Select One or More Answer Choices

According to the *Official Guide to the GRE Revised General Test*, the official directions for "Select One or More Answer Choices" read as follows:

> Directions: Select one or more answer choices according to the specific question directions.
>
> If the question does not specify how many answer choices to select, select all that apply.
>
> The correct answer may be just one of the choices or as many as all of the choices, depending on the question.
>
> No credit is given unless you select all of the correct choices and no others.
>
> If the question specifies how many answer choices to select, select exactly that number of choices.

Note that there is no "partial credit." If three of six choices are correct and you select two of the three, no credit is given. It will also be important to read the directions carefully.

That said, many of these questions look *very* similar to those on the "old" GRE. For instance, here is a question that could have appeared on the GRE in the past:

If $ab = |a| \times |b|$, which of the following *must* be true?

I. $a = b$
II. $a > 0$ and $b > 0$
III. $ab > 0$

A. II only
B. III only
C. I and III only
D. II and III only
E. I, II, and III

Solution: If $ab = |a| \times |b|$, then we know ab is positive, since the right side of the equation must be positive. If ab is positive, however, that doesn't necessarily mean that a and b are each positive; it simply means that they have the same sign.

I. It is not true that a must equal b. For instance, a could be 2 and b could be 3.
II. It is not true that a and b must each be positive. For instance, a could be -3 and b could be -4.
III. True. Since $|a| \times |b|$ must be positive, ab must be positive as well.

The answer is B (III only).

Note that, if you determined that statement I was false, you could eliminate choices C and E before considering the remaining statements. Then, if you were confident that II was also false, you could safely pick answer choice B, III only, without even trying statement III, since "None of the above" isn't an option. That is, because of the multiple choice answers, it is sometimes not necessary to consider each statement individually. This is the aspect of such problems that will change on the new exam.

Here is the same problem, in the new format.

If $ab = |a| \times |b|$, which of the following *must* be true?

Indicate <u>all</u> such statements.

[A] $a = b$
[B] $a > 0$ and $b > 0$
[C] $ab > 0$

Strategy Tip: Make sure to fully "process" the statement in the question (simplify it or list the possible scenarios) before considering the answer choices. This will save you time in the long run!

Here, we would simply select choice C. The only thing that has changed is that we can't do process of elimination; we must always consider each statement individually. On the upside, the problem has become much more straightforward and compact (not every real-life problem has exactly five possible solutions; why should those on the GRE?).

Math: Numeric Entry

This question type requires the text taker to key a numeric answer into a box on the screen. You are not able to "work backwards" from answer choices, and in many cases it will be difficult to make a guess. However, the principles being tested are the same as on the old GRE.

Here is a sample question:

If $x*y = 2xy - (x - y)$, what is the value of 3*4?

Solution:

We are given a function involving two variables, x and y, and asked to substitute 3 for x and 4 for y:

$$x*y = 2xy - (x - y)$$
$$3*4 = 2(3)(4) - (3 - 4)$$
$$3*4 = 24 - (-1)$$
$$3*4 = 25$$

The answer is 25.

Thus, you would type 25 into the box.

Okay. You've now got a good start on understanding the structure and question formats of the new GRE. Now it's time to begin fine-tuning your skills.

Chapter 2
of
FRACTIONS, DECIMALS, & PERCENTS

FRACTIONS

In This Chapter . . .

- Manipulating Fractions
- Switching Between Improper Fractions and Mixed Numbers
- Division in Disguise
- Fraction Operations: Funky Results
- Comparing Fractions: Cross-Multiply
- NEVER Split the Denominator!
- Benchmark Values
- Picking Smart Numbers: Multiples of the Denominators
- When NOT to Use Smart Numbers

FRACTIONS

This chapter is devoted entirely to understanding what fractions are and how they work from the ground up. Let's begin by reviewing the two parts of a fraction: the **numerator** and the **denominator.**

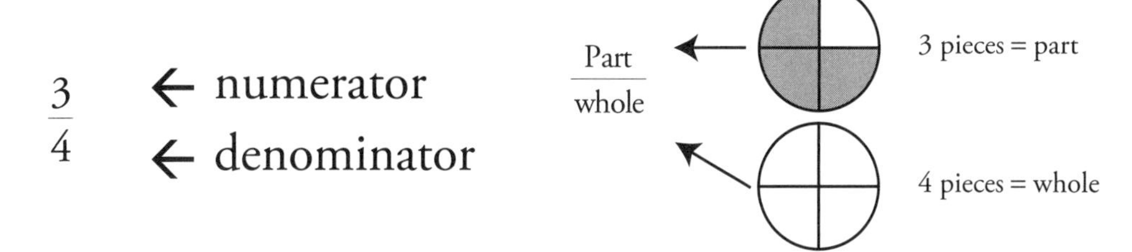

In the picture above, each circle represents a whole unit. One full circle means the number 1, 2 full circles is 2, etc. Fractions essentially divide units into parts. The units above have been divided into 4 equal parts, because the denominator of our fraction is 4. In any fraction, the denominator tells you how many equal pieces a unit has been broken into.

The circle at the top has 3 of the pieces shaded in, and one piece unshaded. That's because the top of our fraction is 3. For any fraction, the numerator tells you how many of the equal pieces you have.

Let's see how changes to the numerator and denominator change a fraction. We'll start by seeing how changes affect the denominator. You've already seen what 3/4 looks like; let's see what 3/5 looks like.

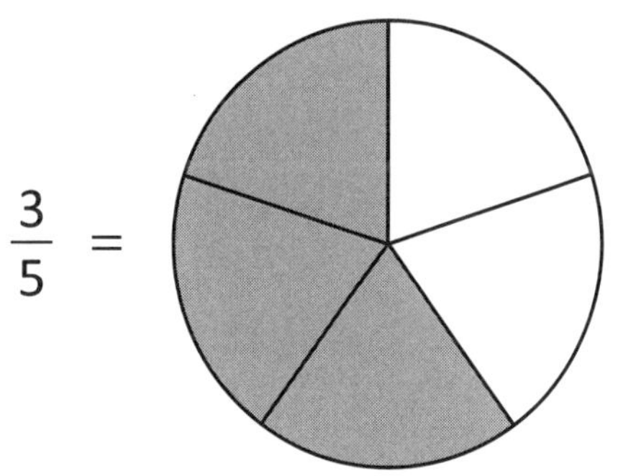

The numerator hasn't changed (it's still 3), so we still have 3 shaded pieces. But now the circle has been divided into 5 pieces instead of 4. One effect is that each piece is now smaller. 1/5 is smaller than 1/4. In general, as the denominator of a number gets bigger, the value of the fraction gets smaller. 3/5 is smaller than 3/4, because each fraction has 3 pieces, but when the circle (or number) is divided into 5 equal portions, each portion is smaller, so 3 portions of 1/5 are less than 3 portions of 1/4.

As we split the circle into more and more pieces, each piece gets smaller and smaller.

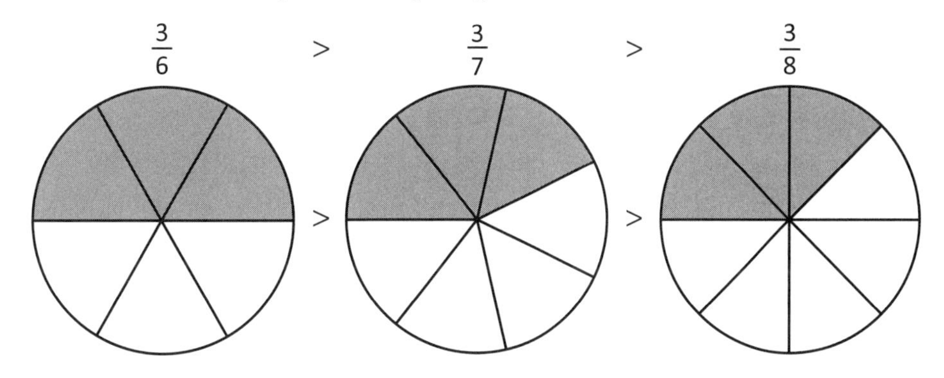

Conversely, as the denominator gets smaller, each piece becomes bigger and bigger.

Now let's see what happens as we change the numerator. The numerator tells us how many pieces we have, so if we make the numerator smaller, we get fewer pieces.

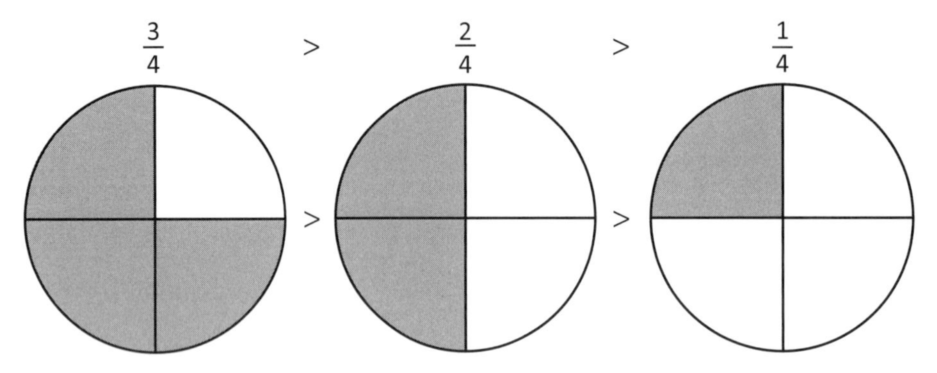

Conversely, if we make the numerator larger, we get more pieces. Let's look more closely at what happens as we get more pieces. In particular, we want to know what happens when the numerator becomes equal to or greater than the denominator. First, let's see what happens when we have the same numerator and denominator. If we have 4/4 pieces, this is what our circle looks like:

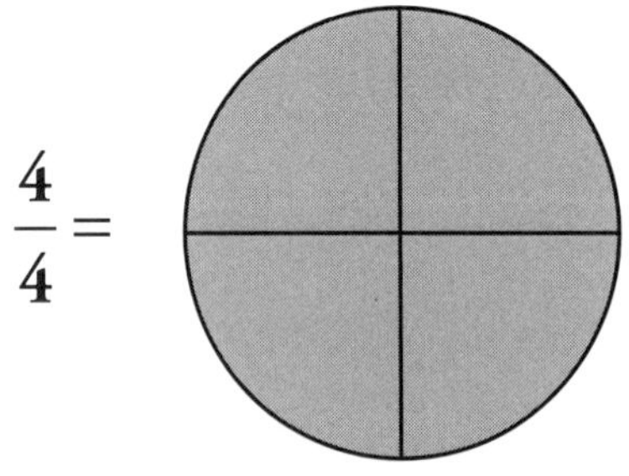

Remember, our circle represents one whole unit. So when all four parts are filled, we have one full unit, or 1. So 4/4 is equal to 1. In general, if the numerator and denominator of a fraction are the same, that fraction equals 1.

Now let's see what happens as the numerator becomes larger than the denominator. What does 5/4 look like?

$$\frac{5}{4} =$$

Each circle is only capable of holding 4 pieces, so when we fill up one circle, we have to move on to a second circle and begin filling it up too. So one way of looking at 5/4 is that we have one complete circle, which we know is equivalent to 1, and we have an additional 1/4. So another way to write 5/4 is 1 + 1/4. This can be shortened to $1\frac{1}{4}$ ("one and one-fourth").

In the last example, the numerator was only a little larger than the denominator. But that will not always be the case. The same logic applies to any situation. Look at the fraction 15/4. Once again, this means that each number is divided into 4 pieces, and we have 15 pieces.

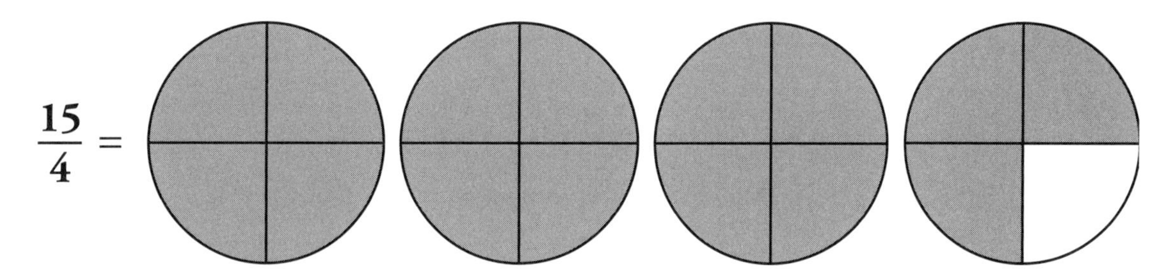

In this case, we have 3 circles completely filled. To fill 3 circles, we needed 12 pieces. (Note: 3 circles × 4 pieces per circle = 12 pieces.) In addition to the 3 full circles, we have 3 additional pieces. So we have $\frac{15}{4} = 3 + \frac{3}{4} = 3\frac{3}{4}$. Whenever you have both an integer and a fraction in the same number, you have a **mixed number** (also called a **proper fraction**). Meanwhile, any fraction in which the numerator is larger than the denominator (for example, 5/4) is known as an **improper fraction.** Improper fractions and mixed numbers express the same thing. Later in the chapter we'll discuss how to change from improper fractions to mixed numbers and vice-versa.

Let's review what we've learned about fractions so far. Every fraction has two components: the numerator and the denominator.

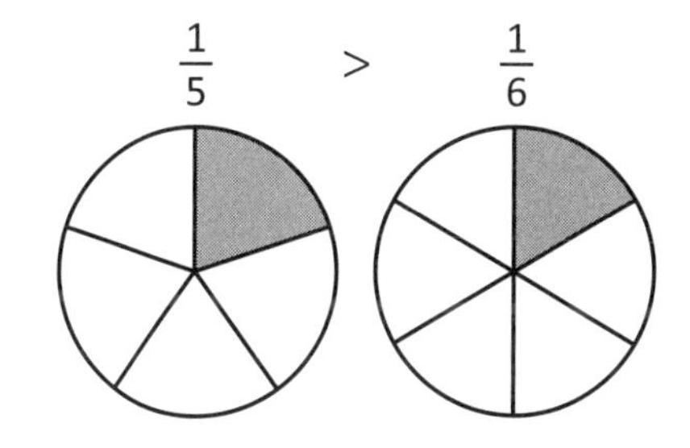

The denominator tells you how many equal pieces each unit circle has. As the denominator gets bigger, each piece gets smaller, so the fraction gets smaller as well.

The numerator tells you how many equal pieces you have. As the numerator gets bigger, you have more pieces, so the fraction gets bigger.

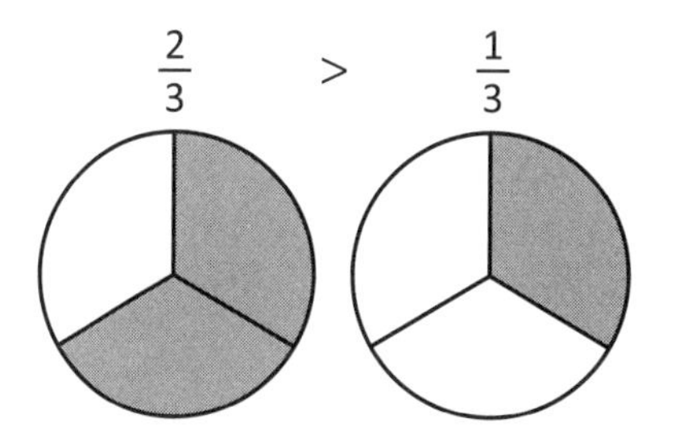

When the numerator is smaller than the denominator, the fraction will be less than 1. When the numerator equals the denominator, the fraction equals 1. When the numerator is larger than the denominator, the fraction is greater than 1.

Check Your Skills

For each of the following sets of fractions, decide which fraction is larger:

1. $\frac{5}{7}$ vs. $\frac{3}{7}$
2. $\frac{3}{10}$ vs. $\frac{3}{13}$

Answers can be found on page 57.

Manipulating Fractions

In the next two sections, we'll discuss how to add, subtract, multiply and divide fractions. We're already familiar with these four basic manipulations of arithmetic, but when fractions enter the picture, things can become more complicated.

Below, we're going to discuss each manipulation in turn. In each discussion, we'll first talk conceptually about what changes are being made with each manipulation. Then we'll go through the actual mechanics of performing the manipulation.

We'll begin with how to add and subtract fractions.

Fraction Addition and Subtraction

The first thing to recall about addition and subtraction in general is that they affect how many things you have. If you have 3 things, and you add 6 more things, you have 3 + 6 = 9 things. If you have 7 things and you subtract 2 of those things, you now have 7 – 2 = 5 things. That same basic principle holds true with fractions as well. What this means is that addition and subtraction affect the numerator of a fraction, because the numerator tells us how many things, or pieces, we have.

For example, let's say we want to add the two fractions 1/5 and 3/5. What we are doing is adding 3 fifths to 1 fifth. (A "fifth" is the very specific pie slice we see below.)

$$\frac{1}{5} + \frac{3}{5}$$

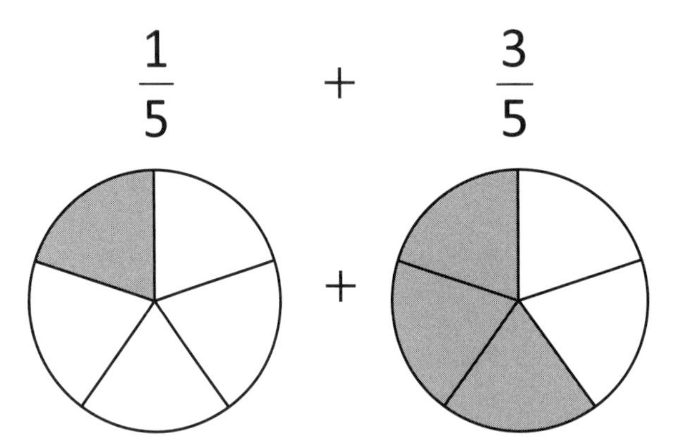

If we were dealing with integers, and we added 3 to 1, we would get 4. The idea is the same with fractions. Now, instead of adding 3 complete units to one complete unit, we're adding 3 fifths to 1 fifth. 1 fifth plus 3 fifths equals 4 fifths.

$$\frac{1}{5} + \frac{3}{5} = \frac{4}{5}$$

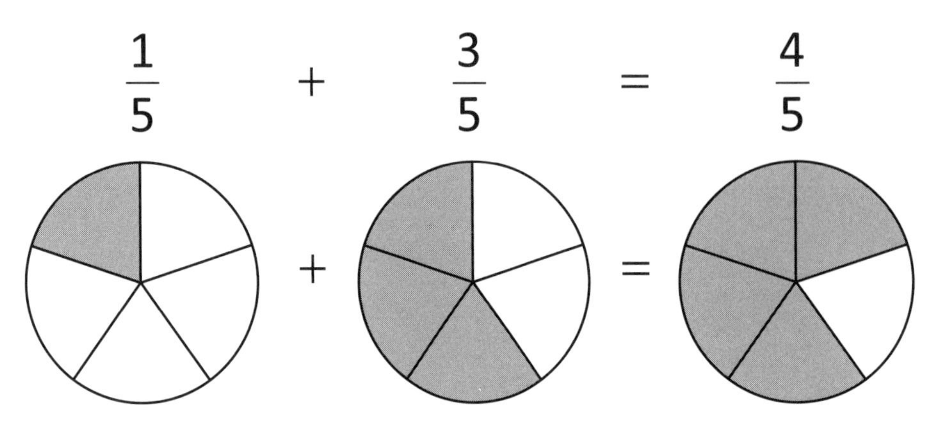

Notice that when we added the two fractions, the denominator stayed the same. Remember, the denominator tells us how many pieces each unit has been broken into. In other words, it determines the size of the slice. Adding 3 pieces to 1 piece did nothing to change the size of the pieces. Each unit is still broken into 5 pieces; hence there is no change to the denominator. The only effect of the addition was to end up with more pieces, which means that we ended up with a larger numerator.

Be able to conceptualize what we just did both ways: adding 1/5 and 3/5 to get 4/5, *and* regarding 4/5 as the sum of 1/5 and 3/5.

$$\frac{1}{5}+\frac{3}{5}=\frac{1+3}{5}=\frac{4}{5}$$

$$\frac{4}{5}=\frac{1+3}{5}=\frac{1}{5}+\frac{3}{5}$$

Also, you should be able to handle an x (or any variable) in place of one of the numerators.

$$\frac{1}{5}+\frac{x}{5}=\frac{4}{5} \text{ becomes } 1+x=4$$

$$x=3$$

We can apply the same thinking no matter what the denominator is. Say we want to add 3/6 and 5/6. This is how it looks:

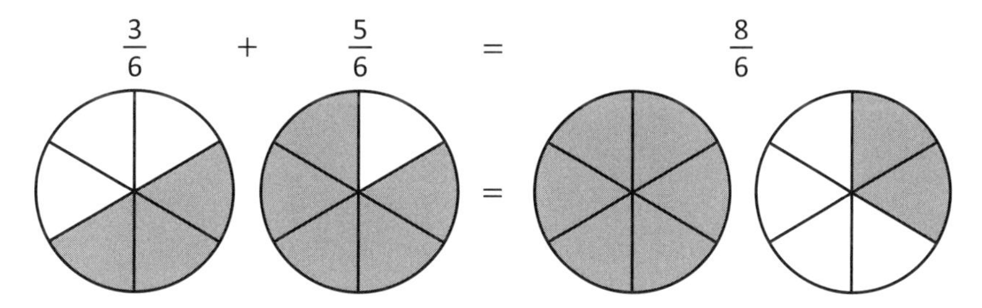

Notice that once again, the only thing that changes during the operation is the numerator. Adding 5 sixths to 3 sixths gives you 8 sixths. The principle is still the same even though we ended up with an improper fraction.

Again, see the operation both ways:

$$\frac{3}{6}+\frac{5}{6}=\frac{3+5}{6}=\frac{8}{6} \qquad \frac{8}{6}=\frac{3+5}{6}=\frac{3}{6}+\frac{5}{6}$$

Be ready for a variable as well:

$$\frac{3}{6}+\frac{x}{6}=\frac{8}{6} \text{ becomes } 3+x=8$$

$$x=5$$

Now let's look at a slightly different problem. This time we want to add 1/4 and 3/8.

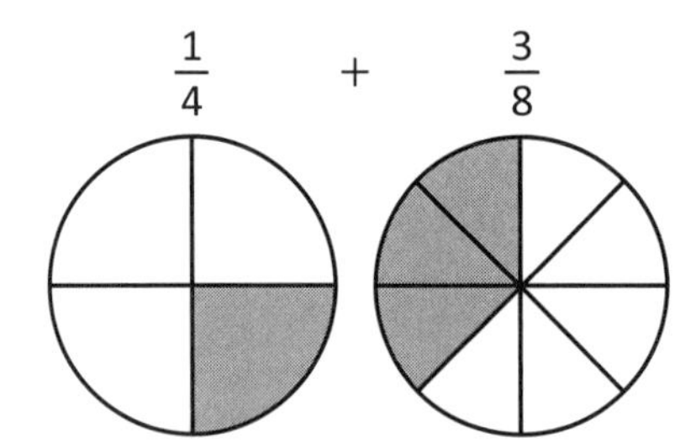

Do you see the problem here? We have one thing on the left and three things on the right, but the denominators are different, so the sizes of the pieces are different. It doesn't make sense in this case simply to add the numerators and get 4 of anything. Fraction addition only works if we can add pieces that are all the same size. So now the question becomes, how can we make all the pieces the same size?

What we need to do is find a new way to express both of the fractions so that the slices are the same size. For this particular addition problem, we can take advantage of the fact that one fourth is twice as big as one eighth. Look what happens if we take all the fourths in our first circle and divide them in two:

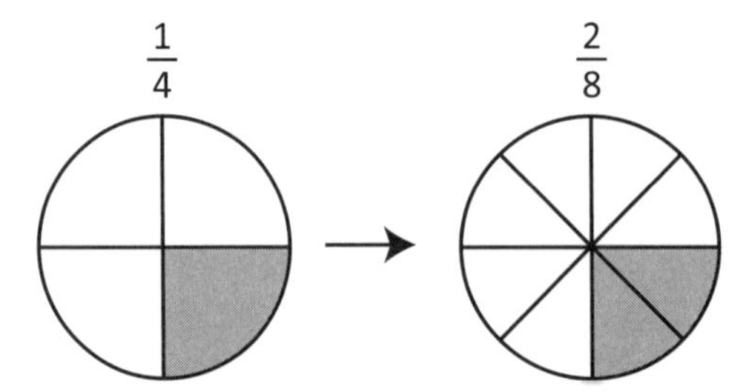

What happened to our fraction? The first thing to note is that we haven't changed the value of our fraction. Originally, we had 1 piece out of 4. Once we divided every part into 2, we ended up with 2 pieces out of 8. So we ended up with twice as many pieces, but each piece was half as big. So we actually ended up with the same amount of "stuff" overall.

What did we change? We ended up with twice as many pieces, which means we multiplied the numerator by 2, and we broke the circle into twice as many pieces, which means we also multiplied the denominator by 2. So we ended up with $\frac{1\times 2}{4\times 2}=\frac{2}{8}$. We'll come back to this concept later, but for now, simply make sure that you understand that $\frac{1}{4}=\frac{2}{8}$.

So without changing the value of 1/4, we've now found a way to *rename* 1/4 as 2/8, so we can add it to 3/8. Now our problem looks like this:

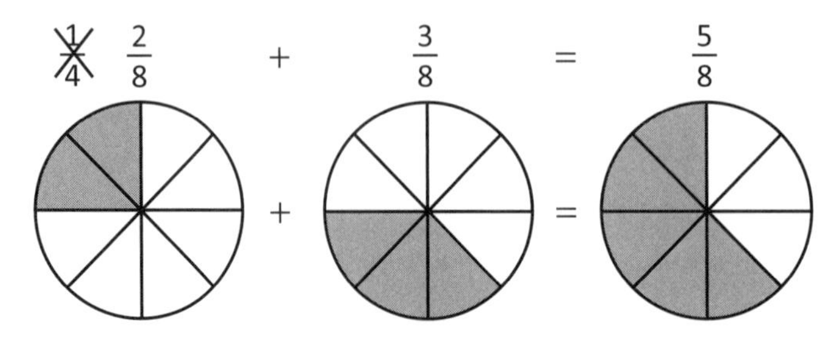

The key to this addition problem was to find what we call a **common denominator.** Finding a common denominator simply means renaming the fractions so they have the same denominator. Then, and *only* then, can we add the renamed fractions.

We won't get into all the details of fraction multiplication just yet (don't worry—it's coming), but we need to take a closer look at what we did to the fraction $\frac{1}{4}$ in order to rename it. Essentially what we did was multiply this fraction by $\frac{2}{2}$. $\frac{1}{4}=\frac{1}{4}\times\frac{2}{2}=\frac{1\times 2}{4\times 2}=\frac{2}{8}$. As we've already discussed, any fraction in which the numerator equals the denominator is 1. So 2/2 = 1. That means that all we did was multiply 1/4 by 1. And anything times 1 equals itself. So we changed the appearance of 1/4 by multiplying the top and bottom by 2, but we did not change its value.

Finding common denominators is a critical skill when dealing with fractions. Let's walk through another example and see how the process works. This time we're going to add 1/4 and 1/3.

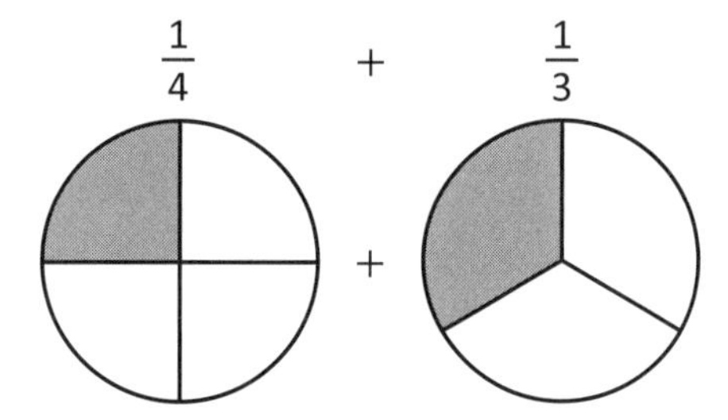

Once again we are adding two fractions with different-sized pieces. There's no way to complete the addition without finding a common denominator. But remember, the only way that we can find common denominators is by multiplying one or both of the fractions by some version of 1 (such as 2/2, 3/3, 4/4, etc.). Because we can only multiply by 1 (the number that won't change the value of the fraction), the only way we can change the denominators is through multiplication. In the last example, the two denominators were 4 and 8. We were able to make them equal because 4 × 2 = 8.

Because all we can do is multiply, what we really need when we look for a common denominator is a common *multiple* of both denominators. In the last example, 8 was a multiple of both 4 and 8.

In this problem, we need to find a number that is a multiple of both 4 *and* 3. List a few multiples of 4: 4, 8, 12, 16. Also list a few multiples of 3: 3, 6, 9, 12, stop. 12 is on both lists, so 12 is a multiple of both 3 and 4. Now we need to change both fractions so that they have a denominator of 12.

Let's begin by changing 1/4. We have to ask the question, what times 4 equals 12? The answer is 3. That means that we want to multiply 1/4 by 3/3. $\frac{1}{4}\times\frac{3}{3}=\frac{3}{12}$. So 1/4 is the same as 3/12. Once again, we can look at our circles to verify these fractions are the same:

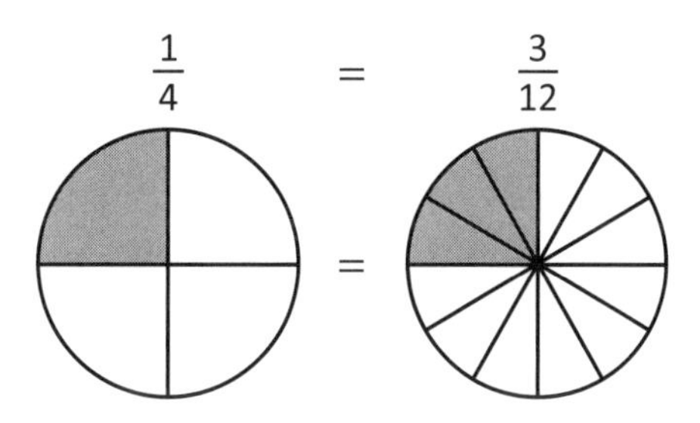

Now we need to change 1/3. We need to ask, what times 3 equals 12? 4 × 3 = 12, so we need to multiply 1/3 by 4/4. $\frac{1}{3}=\frac{1}{3}\times\frac{4}{4}=\frac{4}{12}$. Now both of our fractions have a common denominator, so we're ready to add.

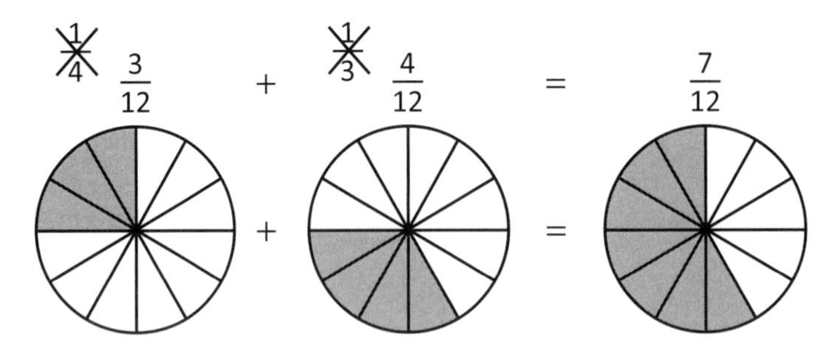

$$\frac{1}{4}+\frac{1}{3}=\frac{1\times 3}{4\times 3}+\frac{1\times 4}{3\times 4}=\frac{3}{12}+\frac{4}{12}=\frac{7}{12}$$

And now you know everything you need to add any two fractions together.

Let's recap what we've done so far:

- When adding fractions, we have to add equal-sized pieces. That means we need the denominators to be the same for any fractions we want to add. **If the denominators are the same, then you add the numerators and keep the denominator the same.**

  Ex. $\frac{2}{9}+\frac{5}{9}=\frac{7}{9}$

- If the two fractions have different denominators, we need to find a common multiple for the two denominators first.

  Ex. $\frac{1}{4}+\frac{2}{5}=?$
  Common multiple of 4 and 5 = 20

- Once we know the common multiple, we need to figure out what number for each fraction multiplies the denominator to reach the common multiple.

  $\frac{1}{4}+\frac{5}{5}=\frac{5}{20}$ $\frac{2}{5}\times\frac{4}{4}=\frac{8}{20}$

- Using the number we found in the last step, we multiply each fraction that needs to be changed by the appropriate fractional version of 1 (such as 5/5).

- Now that the denominators are the same, we can add the fractions.

  $\frac{5}{20}+\frac{8}{20}=\frac{13}{20}$

This section would not be complete without a discussion of subtraction. The good news is that subtraction works exactly the same way as addition! The only difference is that when you subtract, you end up with fewer pieces instead of more pieces, so you end up with a smaller numerator.

Let's walk through a subtraction problem together. What is 5/7 – 1/3?

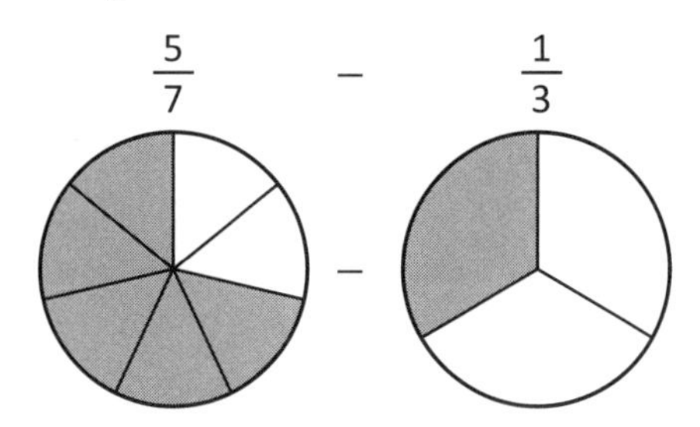

Just like addition, subtraction of fractions requires a common denominator. So we need to figure out a common multiple of the two denominators: 7 and 3. 21 is a common multiple, so let's use that.

Let's change 5/7 so that its denominator is 21. 3 times 7 equals 21, so let's multiply 5/7 by 3/3: $\frac{5}{7}=\frac{5}{7}\times\frac{3}{3}=\frac{15}{21}$. Now we do the same for 1/3. 7 times 3 equals 21, so let's multiply 1/3 by 7/7: $\frac{1}{3}=\frac{1}{3}\times\frac{7}{7}=\frac{7}{21}$. Our subtraction problem can be rewritten as $\frac{15}{21}-\frac{7}{21}$, which we can easily solve.

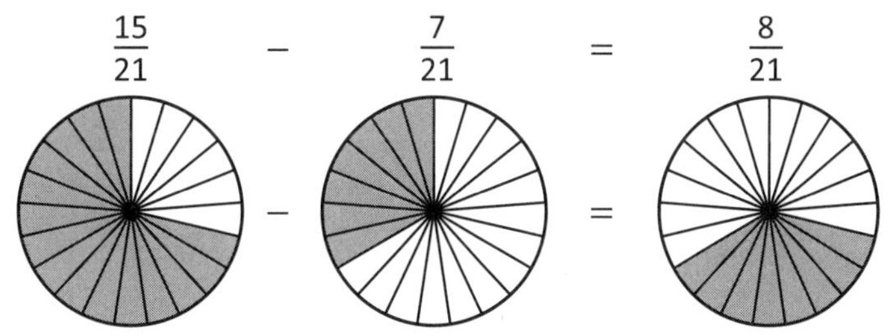

Finally, if you have a variable in the subtraction problem, nothing really changes. One way or another, you still have to find a common denominator.

Let's try another problem.

Solve: $\frac{1}{4}+\frac{x}{5}=\frac{13}{20}$

First, subtract 1/4 from each side:

$$\frac{x}{5}=\frac{13}{20}-\frac{1}{4}$$

Perform the subtraction by finding the common denominator, which is 20.

$$\frac{13}{20}-\frac{1\times5}{4\times5}=\frac{13}{20}-\frac{5}{20}=\frac{8}{20}$$

So we have $\frac{x}{5}=\frac{8}{20}$

There are several options at this point. The one we'll use right now is to convert to the common denominator again (which is still 20).

$$\frac{x\times4}{5\times4}=\frac{4x}{20}=\frac{8}{20}$$

Now we can set the numerators equal: $4x = 8$
Divide by 4: $x = 2$

If we had spotted the common denominator of all 3 fractions at the start, we could have saved work:

$$\frac{1}{4}+\frac{x}{5}=\frac{13}{20}$$

Convert to a common denominator of 20: $\frac{1\times5}{4\times5}+\frac{x\times4}{5\times4}=\frac{13}{20}$

Clean up: $\frac{5}{20}+\frac{4x}{20}=\frac{13}{20}$

Set numerators equal: $5 + 4x = 13$

Subtract 5: $4x = 8$

Divide by 4: $x = 2$

Check Your Skills
Evaluate the following expressions:

3. $\frac{1}{2}+\frac{3}{4}=$

4. $\frac{2}{3}-\frac{3}{8}=$

5. Solve for x. $\frac{x}{5}+\frac{2}{5}=\frac{13}{5}$

6. Solve for x. $\frac{x}{3}-\frac{4}{9}=$

Answers can be found on page 57.

Simplifying Fractions

Suppose you were presented with this question on the GRE.

$\frac{5}{9}+\frac{1}{9}=?$

a. 4/9 b. 5/9 c. 2/3

This question involves fraction addition, which we know how to do. So let's begin by adding the two fractions. $\frac{5}{9}+\frac{1}{9}=\frac{5+1}{9}=\frac{6}{9}$. But 6/9 isn't one of the answer choices. Did we do something wrong? No, we didn't, but we did forget an important step.

6/9 doesn't appear as an answer choice because it isn't simplified (in other words, **reduced**). To understand what that means, we're going to return to a topic that should be very familiar to you at this point: prime factors. Let's break down the numerator and denominator into prime factors. $\frac{6}{9}\rightarrow\frac{2\times3}{3\times3}$.

Notice that both the numerator and the denominator have a 3 as one of their prime factors. Because neither multiplying nor dividing by 1 changes the value of a number, we can effectively cancel the $\frac{3}{3}$, leaving behind only $\frac{2}{3}$. That is, $\frac{6}{9}=\frac{2\times 3}{3\times 3}=\frac{2}{3}\times\frac{\cancel{3}}{\cancel{3}}=\frac{2}{3}$.

Let's look at another example of a fraction that can be reduced: $\frac{18}{60}$. Once again, we can begin by breaking down the numerator and denominator into their respective prime factors. $\frac{18}{60}=\frac{2\times 3\times 3}{2\times 2\times 3\times 5}$. This time, the numerator and the denominator have two factors in common: a 2 and a 3. Once again, we can split this fraction into two pieces:

$$\frac{2\times 3\times 3}{2\times 2\times 3\times 5}=\frac{3}{2\times 5}\times\frac{2\times 3}{2\times 3}=\frac{3}{10}\times\frac{6}{6}$$

Once again, $\frac{6}{6}$ is the same as 1, so really we have $\frac{3}{10}$, which leaves us with $\frac{3}{10}$.

As you practice, you should be able to simplify fractions by recognizing the largest common factor in the numerator and denominator and canceling it out. For example, you should recognize that in the fraction $\frac{18}{60}$, both the numerator and the denominator are divisible by 6. That means we could think of the fraction as $\frac{3\times 6}{10\times 6}$. You can then cancel out the common factors on top and bottom and simplify the fraction:

$$\frac{18}{60}=\frac{3\times\cancel{6}}{10\times\cancel{6}}=\frac{3}{10}.$$

Check Your Skills
Simplify the following fractions.

7. $\frac{25}{40}$

8. $\frac{16}{24}$

Answers can be found on page 57.

Fraction Multiplication

Now that we know how to add and subtract fractions, we're ready to multiply and divide them. We'll begin with multiplication. First, we'll talk about what happens when we multiply a fraction by an integer.

We'll start by asking the question, what is $1/2 \times 6$? When we added and subtracted fractions, we were really adding and subtracting pieces of numbers. With multiplication, conceptually it is different: we are starting with an amount, and leaving a fraction of it behind. For instance, in this example, what we are really asking is, what is 1/2 *of* 6? There are a few ways to conceptualize what that means.

We want to find one half of six. One way to do that is to split 6 into 2 equal parts and keep one of those parts.

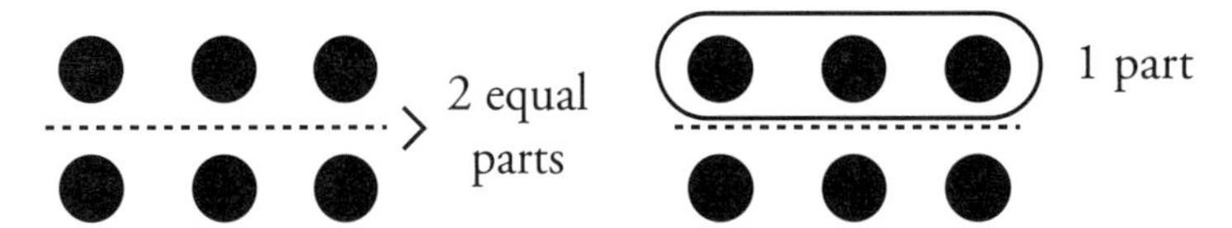

Because the denominator of our fraction is 2, we divide 6 into 2 equal parts of 3. Then, because our denominator is 1, we keep one of those parts. So 1/2 × 6 = 3.

We can also think of this multiplication problem a slightly different way. Consider each unit circle of the 6. What happens if we break each of those circles into 2 parts, and keep 1 part?

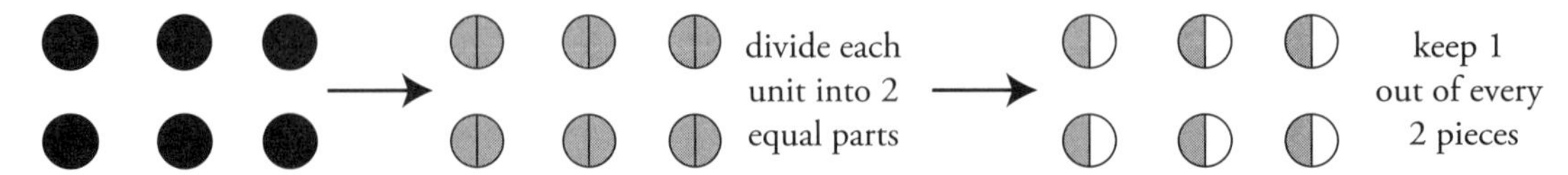

We divide every circle into 2 parts, and keep 1 out of every 2 parts. We end up with 6 halves, or 6/2, written as a fraction. But 6/2 is the same as 3, so really, 1/2 of 6 is 3.

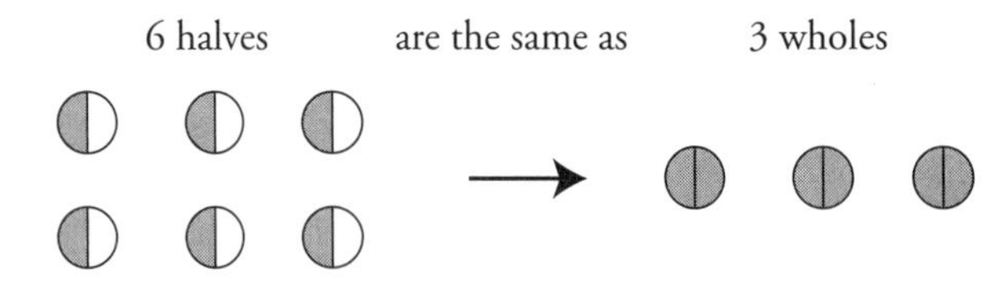

Either way we conceptualize this multiplication, we end up with the same answer. Let's try another example.

What is 2/3 × 12?

Once again, we are really asking, what is 2/3 *of* 12? In the previous example, when we multiplied a number by 1/2, we divided the number into 2 parts (as indicated by the denominator). Then we kept 1 of those parts (as indicated by the numerator).

By the same logic, if we want to get 2/3 of 12, we need to divide 12 into 3 equal parts, because the denominator is 3. Then we keep 2 of those parts, because the numerator is 2. As with the first example, there are several ways of conceptualizing this. One way is to divide 12 into 3 equal parts, and keep 2 of those parts.

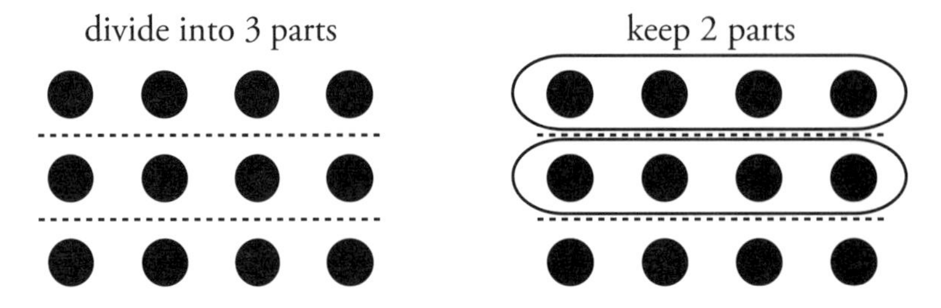

We divided 12 into 3 equal parts of 4, and kept 2 of those parts. 2 groups of 4 is 8, so 2/3 × 12 = 8.

Another way to conceptualize 2/3 × 12 is to once again look at each unit of 12. If we break each unit into 3 pieces (because the denominator of our fraction is 3) and keep 2 out of every 3 pieces (because our numerator is 2) we end up with this:

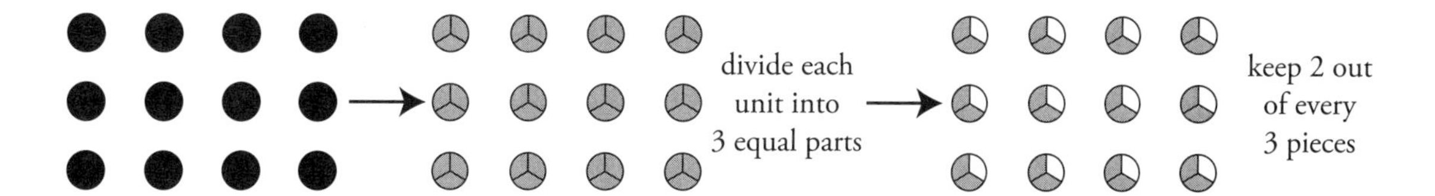

We ended up with 24 thirds, or 24/3. But 24/3 is the same as 8, so 2/3 of 12 is 8.

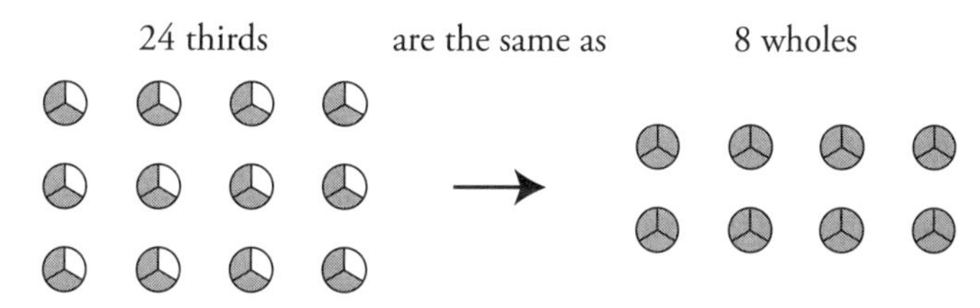

Once again, either way we think about this multiplication problem, we arrive at the same conclusion. $2/3 \times 12 = 8$.

Now that we've seen what happens when we multiply an integer by a fraction, it's time to multiply a fraction by a fraction. It's important to remember that the basic logic is the same. When we multiply any number by a fraction, the denominator of the fraction tells us how many parts to divide the number into, and the numerator tells us how many of those parts to keep. Now let's see how that logic applies to fractions.

What is $\frac{1}{2} \times \frac{3}{4}$?

This question is asking, what is 1/2 *of* 3/4? So once again, we need to divide 3/4 into 2 equal parts. This time, though, because we're splitting a fraction, we're going to do things a little differently. Because 3/4 is a fraction, the unit circle has already broken a number into 4 equal pieces. So what we're going to do is break each of those pieces into 2 smaller pieces.

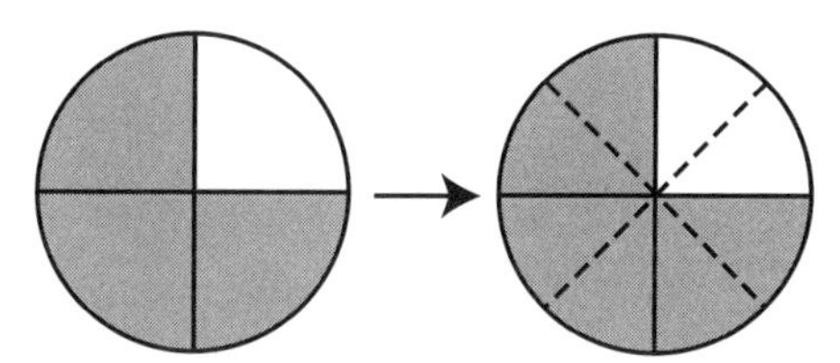

Cut each piece in half

Now that we've divided each piece into 2 smaller pieces, we want to keep 1 of each those smaller pieces.

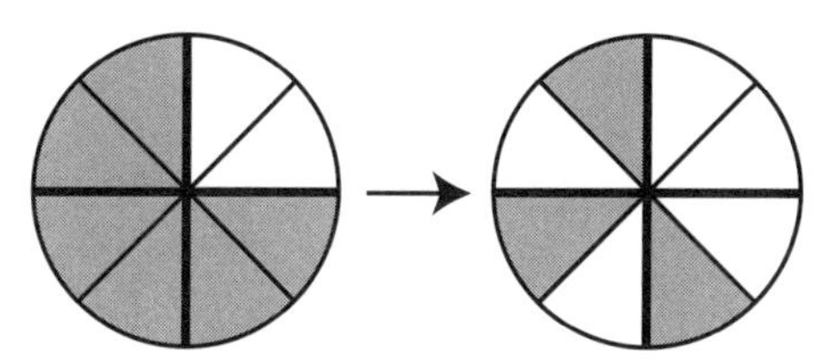

Keep 1 out of each of the 2 resulting pieces.

So what did we end up with? First of all, our result is going to remain a fraction. Our original number was 3/4. In other words, a number was broken into 4 parts, and we kept 3 of those parts. Now the number has been broken into 8 pieces, not 4, so our denominator is now 8. However, we still have $1 \times 3 = 3$ of those parts, so our numerator is still 3. So 1/2 of 3/4 is 3/8.

Let's try one more. What is 5/6 × 1/2? Once again, we have to start by dividing our fraction into 6 equal pieces.

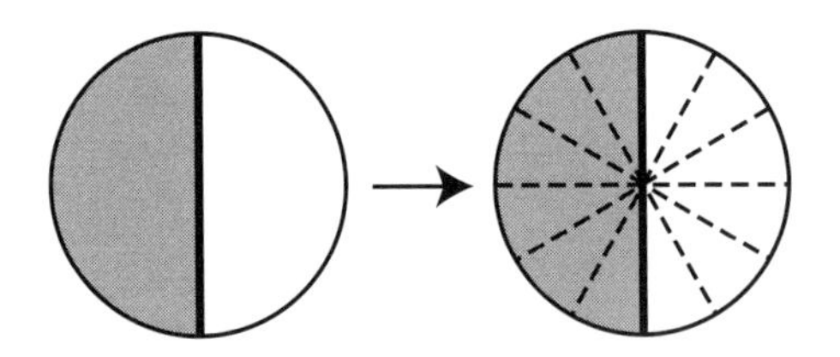

Cut each piece into 6 smaller pieces

Now we want to keep 5 out of every 6 parts.

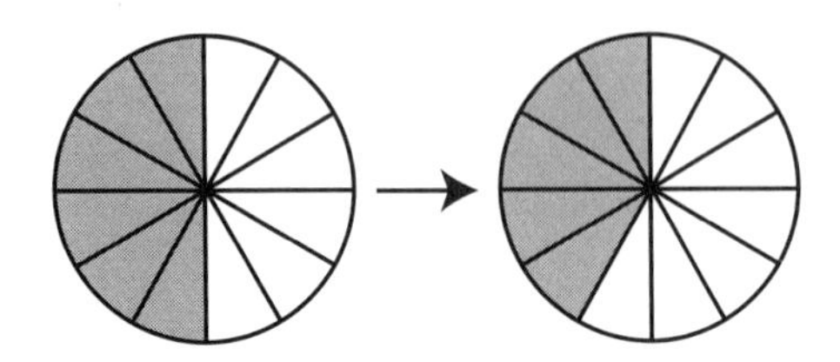

Keep 5 of the 6

So what did we end up with? Now we have a number divided into 12 parts, so the denominator is 12, and we keep 1 × 5 = 5 parts, so the numerator is 5. 5/6 of 1/2 is 5/12.

Multiplying fractions would get very cumbersome if we always resorted to slicing circles up into increasingly tiny pieces. So now, let's talk in a general way about the mechanics of multiplying a number by a fraction.

First, note the following *crucial* difference between two types of arithmetic operations on fractions:

Addition & Subtraction:
Only the numerator changes (once you've found a common denominator).

Multiplication & Division:
Both the numerator and the denominator typically change.

The way to generalize **fraction multiplication** is this: *multiply the numerators together to get the new numerator, and multiply the denominators together to get the new denominator*. Then, simplify (or reduce).

$$\frac{1}{2}\times\frac{6}{1}=\frac{1\times6}{2\times1}=\frac{6}{2}=3$$

$$\frac{2}{3}\times\frac{12}{1}=\frac{2\times12}{3\times1}=\frac{24}{3}=8$$

$$\frac{1}{2}\times\frac{3}{4}=\frac{1\times3}{2\times4}=\frac{3}{8}$$

$$\frac{5}{6}\times\frac{1}{2}=\frac{5\times1}{6\times2}=\frac{5}{12}$$

In practice, when we multiply fractions, *we don't need to worry about the conceptual foundation* once we understand the mechanics.

$\frac{1}{2}\times\frac{3}{4}$ → Mechanics: $\frac{1}{2}\times\frac{3}{4}=\frac{1\times 3}{2\times 4}=\frac{3}{8}$ EASIER

$\frac{1}{2}\times\frac{3}{4}$ → Conceptual: "one half of 3/8" ...cut up the circles further... HARDER

Finally, whenever we multiply fractions, we should always look to cancel common factors, in order to reduce our answer without doing unnecessary work.

$$\frac{33}{7}\times\frac{14}{3}=?$$

The long way to do this is:

$33 \times 14 =$

$$\begin{array}{r} {}^{1}33 \\ \times\ 14 \\ \hline 132 \\ 330 \\ \hline 462 \end{array}$$

$7 \times 3 = 21$

You wind up with $\frac{462}{21}$:

$$\begin{array}{r} 22 \\ 21\overline{)462} \\ -42 \\ \hline 42 \\ -42 \\ \hline 0 \end{array}$$

This work can be simplified greatly by *canceling* parts of each fraction *before* multiplying. Always look for common factors in the numerator and denominator:

$$\frac{33}{7}\times\frac{14}{3}=\frac{3\times 11}{7}\times\frac{2\times 7}{3}$$

We can now see that the numerator of the first fraction has a 3 as a factor, which can be canceled out with the 3 in the denominator of the second fraction. (This is because multiplication and division operate at the same level of priority in the PEMDAS operations!) Similarly, the 7 in the denominator of the first fraction can be canceled out by the 7 in the numerator of the second fraction. By cross-canceling these factors, we can save ourselves a lot of work.

$$\frac{\cancel{3}\times 11}{\cancel{7}}\times\frac{2\times\cancel{7}}{\cancel{3}}=\frac{11}{1}\times\frac{2}{1}=\frac{22}{1}=22$$

Check Your Skills

Evaluate the following expressions. Simplify all fractions:

9. $\frac{3}{10}\times\frac{6}{7}=$

10. $\frac{5}{14}\times\frac{7}{20}=$

Answers can be found on page 57.

Fraction Division

Now we're up to the last of the four basic arithmetic operations on fractions (addition, subtraction, multiplication and division). This section will be a little different than the other three—it will be different because we're actually going to do fraction division by avoiding division altogether!

We can avoid division entirely because of the relationship between multiplication and division. Multiplication and division are two sides of the same coin. Any multiplication problem can be expressed as a division problem, and vice-versa. This is useful because, although the mechanics for multiplication are straightforward, the mechanics for division are more *work* and therefore more *difficult.* Thus, we are going to express every fraction division problem as a fraction multiplication problem.

Now the question becomes: how do we rephrase a division problem so that it becomes a multiplication problem? The key is **reciprocals.**

Reciprocals are numbers that, when multiplied together, equal 1. For instance, 3/5 and 5/3 are reciprocals, because $\frac{3}{5}\times\frac{5}{3}=\frac{3\times5}{5\times3}=\frac{15}{15}=1$.

Another pair of reciprocals is 2 and 1/2, because $2\times\frac{1}{2}=\frac{2}{1}\times\frac{1}{2}=\frac{2\times1}{1\times2}=\frac{2}{2}=1$. (Once again, it is important to remember that *every integer can be thought of as a fraction.*)

The way to find the reciprocal of a number turns out to generally be very easy—take the numerator and denominator of a number, and switch them around:

	Fraction	Reciprocal
	$\frac{3}{5}$	$\frac{5}{3}$
Integer $2=$	$\frac{2}{1}$	$\frac{1}{2}$

Reciprocals are important because *dividing by a number is the exact same thing as multiplying by its reciprocal*. Let's look at an example to clarify:

What is $6 \div 2$?

This problem shouldn't give you any trouble—6 divided by 2 is 3. But it should also seem familiar: it's the exact same problem we dealt with in the discussion on fraction multiplication. $6 \div 2$ is the *exact same thing* as $6 \times 1/2$.

$6 \div 2 = 3$
$6 \times 1/2 = 3$ — Dividing by 2 is the same as multiplying by 1/2.

To change from division to multiplication, you need to do two things. First, take the divisor (the number to the right of the division sign—in other words, what you are dividing *by*) and replace it with its reciprocal. In this problem, 2 is the divisor, and 1/2 is the reciprocal of 2. Then, switch the division sign to a multiplication sign. So $6 \div 2$ becomes $6 \times 1/2$. Then, proceed to do the multiplication.

$$6 \div 2 = 6 \times \frac{1}{2} = \frac{6}{1} \times \frac{1}{2} = \frac{6 \times 1}{1 \times 2} = \frac{6}{2} = 3$$

This is obviously overkill for $6 \div 2$, but let's try another one. What is $5/6 \div 4/7$?

Once again, we start by taking the divisor (4/7) and replacing it with its reciprocal (7/4). We then switch the division sign to a multiplication sign. So $5/6 \div 4/7$ is the same as $5/6 \times 7/4$. Now we do fraction multiplication:

$$\frac{5}{6} \div \frac{4}{7} = \frac{5}{6} \times \frac{7}{4} = \frac{5 \times 7}{6 \times 4} = \frac{35}{24}.$$

Note that the fraction bar (sometimes indicated with a slash) is another way to express division. After all, $6 \div 2 = 6/2 = \frac{6}{2} = 3$. In fact, the division sign, $\div$, looks like a little fraction. So if you see a "double-decker" fraction, don't worry. It's just one fraction divided by another fraction.

$$\frac{\frac{5}{6}}{\frac{4}{7}} = \frac{5}{6} \div \frac{4}{7} = \frac{5}{6} \times \frac{7}{4} = \frac{35}{24}$$

To recap:

- When you are confronted with a division problem involving fractions, it is *always* easier to perform multiplication than division. For that reason, every fraction division problem should be rewritten as a multiplication problem.

- To do so, replace the divisor with its reciprocal. To find the reciprocal of a number, we simply need to switch the numerator and denominator (ex. 2/9 → 9/2).

Fraction		Reciprocal
$\frac{2}{9}$	→	$\frac{9}{2}$

- Remember that a number multiplied by its reciprocal equals 1.

$$\frac{9}{2} \times \frac{2}{9} = 1$$

- After that, switch the division symbol to a multiplication symbol, and perform fraction multiplication.

$$\frac{3}{4} \div \frac{2}{9} \rightarrow \frac{3}{4} \times \frac{9}{2} = \frac{27}{8}$$

Check Your Skills

Evaluate the following expressions. Simplify all fractions.

11. $\frac{1}{6} \div \frac{1}{11} =$

12. $\frac{8}{5} \div \frac{4}{15} =$

Answers can be found on page 58.

Fractions in Equations

When an x appears in a fraction multiplication or division problem, we'll use essentially the same concepts and techniques to solve.

$$\frac{4}{3}x = \frac{15}{8}$$

Divide both sides by $\frac{4}{3}$: $x = \frac{15}{8} \div \frac{4}{3}$

$$x = \frac{15}{8} \times \frac{3}{4} = \frac{45}{32}$$

An important tool to add to our arsenal at this point is *cross-multiplication.* This tool comes from the principle of making common denominators.

$$\frac{x}{7} = \frac{5}{8}$$

The common denominator of 7 and 8 is 7 × 8 = 56. So we have to multiply the left fraction by 8/8 and the right fraction by 7/7:

$$\frac{8 \times x}{8 \times 7} = \frac{5 \times 7}{8 \times 7} \rightarrow \frac{8x}{56} = \frac{35}{56}$$

Now we can set the numerators equal: $8x = 5 \times 7 = 35$

$x = 35/8$

However, in this situation we can avoid having to determine the common denominator by cross–multiplying each numerator times the other denominator and setting the products equal to each other.

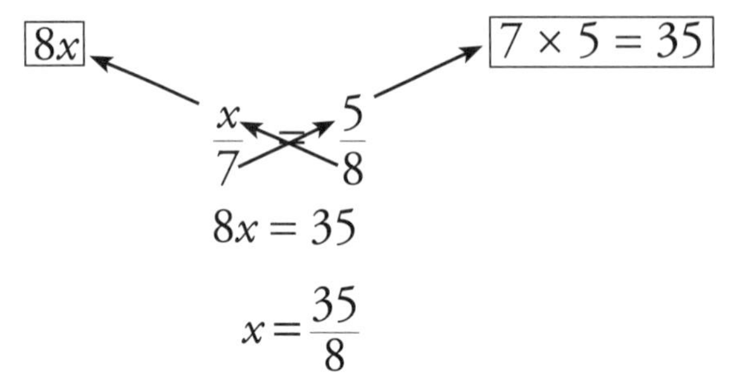

We will further discuss the very important technique of cross-multiplying later in this chapter.

Check Your Skills

Solve for x in the following equations:

13. $\frac{3}{4}x = \frac{3}{2}$

14. $\frac{x}{6} = \frac{5}{3}$

Answers can be found on page 58.

Switching Between Improper Fractions and Mixed Numbers

Let's return to our discussion of why 5/4 equals $1^1/4$ and how to switch between improper fractions and mixed numbers (also known as proper fractions).

To do this we need to talk about the numerator in more detail. The numerator is a description of how many parts we have. The fraction 5/4 tells us that we have 5 parts. But we have some flexibility in how we arrange those 5 parts. For instance, we already expressed it as 4/4 + 1/4, or 1 + 1/4. Essentially what we did was to split the numerator into two pieces: 4 and 1. If we wanted to express this as a fraction, we could say that 5/4 becomes $\frac{4+1}{4}$. This hasn't changed anything, because 4 + 1 equals 5, so we still have the same number of parts.

Then, as we saw above, we can split our fraction into two separate fractions. For instance $\frac{4+1}{4}$ becomes $\frac{4}{4}+\frac{1}{4}$. This is the same as saying that 5 fourths equals 4 fourths plus 1 fourth. So we have several different ways of representing the same fraction. $\frac{5}{4}=\frac{4+1}{4}=\frac{4}{4}+\frac{1}{4}$. Here is a visual representation:

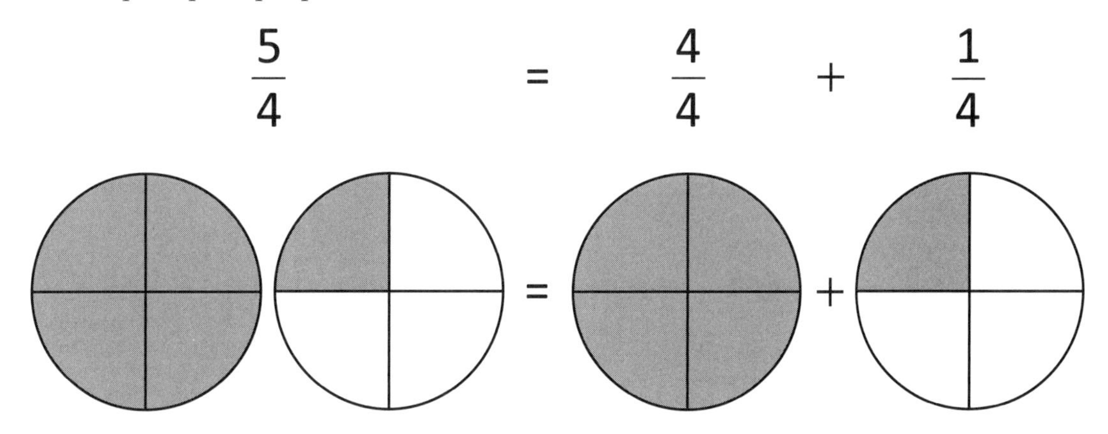

As a general rule, *we can always split the numerator of a fraction into different parts and thus split a fraction into multiple fractions*. This is just *reversing* the process of adding fractions. When we add fractions, we take two fractions with the same denominator and combine them into one fraction. Here we are doing the exact opposite—turning one fraction into two separate fractions, each with the same denominator. And now that our fraction 5/4 is split into two fractions, we can take advantage of the fact that fractions, at their essence, represent *division*: as we discussed earlier, 4/4 = 1 and another way to think of 4/4 is 4 ÷ 4.

To switch from an improper fraction to a mixed number, we want to figure out how many complete units we have. To do that, we need to figure out the *largest multiple of the denominator that is less than or equal to the numerator*. For the fraction 5/4, 4 is the largest multiple of 4 that is less than 5. So we split our fraction into 4/4 and 1/4. We then note that 4/4 equals 1, so our mixed number is $1^1/_4$.

Let's try it again with the fraction 15/4. This time, the largest multiple of 4 that is less than 15 is 12. So we can split our fraction 15/4 into 12/4 + 3/4. In other words, $\frac{15}{4}=\frac{12+3}{4}=\frac{12}{4}+\frac{3}{4}$. And 12/4 = 3, so the fraction 15/4 becomes the mixed number $3\,^3/_4$.

Let's try one with a different denominator. How do we turn the fraction 16/7 into a mixed number? This time we need the largest multiple of 7 that is less than or equal to 16. 14 is the largest multiple of 7 that is less than 16, so we once again split our fraction 16/7 into 14/7 and 2/7.

$\frac{16}{7}=\frac{14+2}{7}=\frac{14}{7}+\frac{2}{7}$. 14 divided by 7 equals 2, so our mixed number is $2^2/_7$.

Check Your Skills
Change the following improper fractions to mixed numbers:

15. $\frac{11}{6}$

16. $\frac{100}{11}$

Answers can be found on page 58.

Changing Mixed Numbers to Improper Fractions

Now that we know how to change a number from an improper fraction to a mixed number (or proper fraction), we also need to be able to do the reverse. Suppose we have the mixed number $5\,^2/_3$. How do we turn this number into a fraction?

Remember that we can think of any integer as a fraction. The number 1, for instance, can be thought of any number of different ways. It can be thought of as 1/1. It can also be thought of as 2/2. In other words, a unit circle can be split into 2 equal pieces, with 2 of those pieces forming a whole unit circle again. 1 can also be written as 3/3, 4/4, 5/5, etc.

In fact, we can think of the process of turning mixed numbers into improper fractions as simple fraction addition. $5\,^2/_3$ is the same thing as 5 + 2/3, so we can think of it as $\frac{5}{1}+\frac{2}{3}$. Now we know what to do—we need to change $\frac{5}{1}$ so that it has a denominator of 3. The way to do that is to multiply $\frac{5}{1}$ by $\frac{3}{3}$. $5=\frac{5}{1}=\frac{5}{1}\times\frac{3}{3}=\frac{5\times3}{1\times3}=\frac{15}{3}$. So our mixed number is really $\frac{15}{3}+\frac{2}{3}=\frac{15+2}{3}=\frac{17}{3}$.

Check Your Skills
Change the following mixed numbers to improper fractions.

17. $3\frac{3}{4}$

18. $5\frac{2}{3}$

Answers can be found on page 58.

Division in Disguise

Sometimes, dividing fractions can be written in a confusing way. Consider one of the previous examples:

$\frac{1}{2} \div \frac{3}{4}$ can also be written as a "double–decker," or **complex**, fraction like this: $\dfrac{\frac{1}{2}}{\frac{3}{4}}$

Do not be confused. You can rewrite this as the top fraction divided by the bottom fraction, and solve it normally (by using the reciprocal of the bottom fraction and then multiplying).

$$\dfrac{\frac{1}{2}}{\frac{3}{4}} = \frac{1}{2} \div \frac{3}{4} = \frac{1}{2} \times \frac{4}{3} = \frac{4}{6} = \frac{2}{3}$$

Here's a **speed tip** for problems like this: notice that, quite often, we can quickly simplify by multiplying both the top fraction and the bottom fraction by a common denominator:

$$\dfrac{\frac{1}{2}}{\frac{3}{4}} = \dfrac{\frac{1}{2} \times 4}{\frac{3}{4} \times 4} = \frac{2}{3}$$

Check Your Skills

Evaluate the following complex fractions by multiplying the top and bottom fractions by a common denominator:

19. $\dfrac{\frac{3}{5}}{\frac{2}{3}} = ?$

20. $\dfrac{\frac{5}{7}}{\frac{1}{4}} = ?$

Answers can be found on page 58.

Fraction Operations: Funky Results

Adding and subtracting fractions leads to expected results: when we add two positive fractions, we get a larger number; when we subtract a positive fraction from something else, we get a smaller number.

However, multiplying by fractions between 0 and 1 yields UNEXPECTED results:

$9 \times \frac{1}{3} = 3$ $\qquad$ $3 < 9$

Multiplying a number by a fraction between 0 and 1 creates a product SMALLER than the original number. Note that this is also true when the original number is a fraction:

$$\frac{1}{2}\times\frac{1}{4}=\frac{1}{8} \qquad \frac{1}{8}<\frac{1}{2}$$

Similarly, dividing by a fraction between 0 and 1 yields a quotient, or result, that is LARGER than the original number:

$$\frac{6}{\frac{3}{4}}=6\div\frac{3}{4}=6\times\frac{4}{3}=\frac{24}{3}=8 \qquad 8>6$$

This is also true when the original number is a fraction:

$$\frac{\frac{1}{4}}{\frac{5}{6}}=\frac{1}{4}\div\frac{5}{6}=\frac{1}{4}\times\frac{6}{5}=\frac{6}{20}=\frac{3}{10} \qquad \frac{3}{10}>\frac{1}{4}$$

Check Your Skills

21. $\frac{1}{2}\times\frac{1}{4}=?$

22. $\frac{1}{2}\div\frac{1}{4}=?$

Answers can be found on page 58.

Comparing Fractions: Cross-Multiply

Earlier we were introduced to the technique of *cross-multiplying* in the context of solving for a variable in an equation that involved fractions. Now let's look at another use of cross-multiplication:

Which fraction is greater, $\frac{7}{9}$ or $\frac{4}{5}$?

The traditional technique used to compare fractions involves finding a common denominator, multiplying, and comparing the two fractions. The common denominator of 9 and 5 is 45.

Thus, $\frac{7}{9}=\frac{35}{45}$ and $\frac{4}{5}=\frac{36}{45}$. Since 35 < 36, we can see that $\frac{4}{5}$ is slightly bigger than $\frac{7}{9}$.

However, there is a shortcut to comparing fractions called (you guessed it): **cross–multiplication**. This is a process that involves multiplying the numerator of one fraction with the denominator of the other fraction, and vice versa. Here's how it works:

$\frac{7}{9}$ $\quad$ $\frac{4}{5}$ — Set up the fractions next to each other.

(7 × 5) $\quad$ (4 × 9)
$\frac{7}{9}$ ⤫ $\frac{4}{5}$ — Cross–multiply the fractions and put each answer by the corresponding *numerator* (**NOT** the denominator!)

35 $<$ 36 — Since 35 is less than 36, the first fraction must be less than the second one.

Check Your Skills

23. Which fraction is greater? $\frac{4}{13}$ or $\frac{1}{3}$?

24. Which fraction is smaller? $\frac{5}{9}$ or $\frac{7}{13}$?

Answers can be found on pages 58–59.

NEVER Split the Denominator!

One final rule—perhaps the most important—is one that you must *always* remember when working with fractions that have an expression (more than one term) in the numerator or denominator. Three examples are:

(a) $\frac{15+10}{5}$ $\qquad$ (b) $\frac{5}{15+10}$ $\qquad$ (c) $\frac{15+10}{5+2}$

In example (a), the numerator is expressed as a sum.
In example (b), the denominator is expressed as a sum.
In example (c), both the numerator and the denominator are expressed as sums.

When simplifying fractions that incorporate sums (or differences), remember this rule: You may split up the terms of the numerator, *but you may NEVER split the terms of the DENOMINATOR*.

Thus, the terms in example (a) may be split:

$$\frac{15+10}{5}=\frac{15}{5}+\frac{10}{5}=3+2=5$$

But the terms in example (b) *may not* be split:

$$\frac{5}{15+10}\neq\frac{5}{15}+\frac{5}{10}\quad \textbf{NO!}$$

Instead, simplify the denominator first:

$$\frac{5}{15+10}=\frac{5}{25}=\frac{1}{5}$$

The terms in example (c) may not be split either:

$$\frac{15+10}{5+2} \neq \frac{15}{5} + \frac{10}{2} \quad \textbf{NO!}$$

Instead, simplify both parts of the fraction:

$$\frac{15+10}{5+2} = \frac{25}{7} = 3\frac{4}{7}$$

Often, GRE problems will involve complex fractions with variables. On these problems, it is tempting to split the denominator. ***Do not fall for it!***

It is tempting to perform the following simplification:

$$\frac{5x-2y}{x-y} = \frac{5x}{x} - \frac{2y}{y} = 5-2=3 \quad \textbf{NO!}$$

This is **WRONG** because you cannot split terms in the denominator.

The reality is that $\frac{5x-2y}{x-y}$ *cannot be simplified further*.

On the other hand, the expression $\frac{6x-15y}{10}$ can be simplified by splitting the difference, *because this difference appears in the numerator*:

Thus: $\frac{6x-15y}{10} = \frac{6x}{10} - \frac{15y}{10} = \frac{3x}{5} - \frac{3y}{2}$

Check Your Skills

Simplify the following fractions:

25. $\frac{13+7}{5}$

26. $\frac{21+6}{7+6}$

27. $\frac{48a+12b}{a+b}$

28. $\frac{9g-6h}{6g-4h}$

Answers can be found on page 59.

Benchmark Values

You will use a variety of estimating strategies on the GRE. One important strategy for estimating with fractions is to use **Benchmark Values**. These are simple fractions with which you are already familiar:

$$\frac{1}{10}, \frac{1}{5}, \frac{1}{4}, \frac{1}{3}, \frac{1}{2}, \frac{2}{3}, \text{ and } \frac{3}{4}$$

You can use Benchmark Values to compare fractions:

Which is greater: $\frac{127}{255}$ or $\frac{162}{320}$?

If we recognize that 127 is <u>less</u> than half of 255, and 162 is <u>more</u> than half of 320, we will save ourselves a lot of cumbersome computation.

You can also use Benchmark Values to estimate computations involving fractions:

Approximately what is $\frac{10}{22}$ of $\frac{5}{18}$ of 2,000?

If we recognize that these fractions are very close to the Benchmark Values $\frac{1}{2}$ and $\frac{1}{4}$, we can estimate:

$\frac{1}{2}$ of $\frac{1}{4}$ of $2{,}000 = \frac{1}{2} \times \frac{1}{4} \times 2{,}000 = 250$. Therefore, $\frac{10}{22}$ of $\frac{5}{18}$ of $2{,}000 \approx 250$.

Notice that the rounding errors compensated for each other:

$\frac{10}{22} \approx \frac{10}{20} = \frac{1}{2}$ We decreased the denominator, so we rounded up: $\frac{10}{22} < \frac{1}{2}$

$\frac{5}{18} \approx \frac{5}{20} = \frac{1}{4}$ We increased the denominator, so we rounded down: $\frac{5}{18} > \frac{1}{4}$

Note also that $\frac{10}{22} \times \frac{5}{18} \times 2{,}000 = \frac{100{,}000}{396} = \frac{25{,}000}{99} = 252.\overline{525}$, so our estimation was very close.

If instead we had rounded $\frac{5}{18}$ to $\frac{6}{18} = \frac{1}{3}$ instead, then you would have rounded *both* fractions up. This would lead to a *slight* but *systematic* overestimation:

$$\frac{1}{2} \times \frac{1}{3} \times 2000 \approx 333$$

<u>*Try to make your rounding errors partially cancel each other out by rounding some numbers up and others down.*</u>

Check Your Skills

29. Which is greater: $\frac{123}{250}$ or $\frac{171}{340}$?

30. Approximate $\left(\frac{15}{58}\right)\left(\frac{9}{19}\right)403$

Answers can be found on page 59.

Picking Smart Numbers: Multiples of the Denominators

Sometimes, fraction problems on the GRE include **unspecified** numerical amounts; sometimes these unspecified amounts are described by variables; other times they are not. In these cases, we often can pick **real numbers** to stand in for the variables. To make the computation easier, choose **Smart Numbers** equal to *common multiples of the denominators of the fractions in the problem*.

For example, consider this problem:

> The Crandalls' hot tub is halfway filled. Their swimming pool, which has a capacity four times that of the hot tub, is filled to four-fifths of its capacity. If the hot tub is drained into the swimming pool, to what fraction of its capacity will the swimming pool be filled?

The denominators in this problem are 2 (from 1/2 of the hot tub) and 5 (from 4/5 of the swimming pool). The Smart Number in this case is the least common denominator, which is 10. Therefore, assign the hot tub, *the smaller quantity*, a capacity of 10 units. Since the swimming pool has a capacity 4 times that of the hot tub, the swimming pool has a capacity of 40 units. We know that the hot tub is only halfway filled; therefore, it has 5 units of water in it. The swimming pool is four-fifths of the way filled, so it has 32 units of water in it.

Let us add the 5 units of water from the hot tub to the 32 units of water that are already in the swimming pool: 32 + 5 = 37.

With 37 units of water and a total capacity of 40, the pool will be filled to $\frac{37}{40}$ of its total capacity.

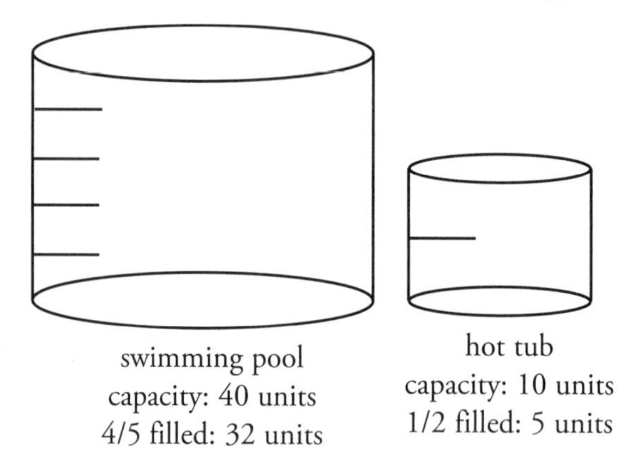

Check Your Skills

Choose Smart Numbers to solve the following problem:

31. Mili's first-generation uHear is filled to 1/2 capacity with songs. Her second-generation uHear, which has 3 times the capacity of her first-generation uHear, is filled to 4/5 capacity. Will Mili be able to transfer all of her music from her first-generation uHear to her second-generation uHear?

Answers can be found on page 59.

When NOT to Use Smart Numbers

In some problems, even though an amount might be unknown to you, it is actually specified in the problem in another way—specifically, because some other related quantity *is given*. In these cases, you ***cannot use*** Smart Numbers to assign real numbers to the variables. For example, consider this problem:

> Mark's comic book collection contains 1/3 Killer Fish comics and 3/8 Shazaam Woman comics. The remainder of his collection consists of Boom comics. If Mark has 70 Boom comics, how many comics does he have in his entire collection?

Even though you do not know the number of comics in Mark's collection, you can see that the total is not completely unspecified. You know a *piece* of the total: 70 Boom comics. You can use this information to find the total. *Do not use Smart Numbers here*. Instead, solve problems like this one by figuring out how big the known piece is; then, use that knowledge to find the size of the whole. You will need to set up an equation and solve:

$$\frac{1}{3}\text{ Killer Fish} + \frac{3}{8}\text{ Shazaam Woman} = \frac{17}{24}\text{ comics that are not Boom}$$

Therefore, $\frac{24}{24} - \frac{17}{24} = \frac{7}{24}$ of the comics are in fact Boom comics.

$$\frac{7}{24}x = 70$$

$$x = 70 \times \frac{24}{7}$$

$$x = 240$$

Thus, Mark has 240 comics.

In summary, **do** pick smart numbers when *no amounts* are given in the problem, but **do not** pick smart numbers when *any amount or total* is given!

Check Your Skills

DO NOT choose Smart Numbers to solve the following problem.

32. John spends 1/3 of his waking hours working, 1/5 of his waking hours eating meals, 3/10 of his waking hours at the gym, and 2 hours going to and from work. He engages in no other activities while awake. How many hours is John awake?

Answers can be found on page 60.

Check Your Skills Answers

1. $\mathbf{\frac{5}{7}}$**:** The denominators of the two fractions are the same, but the numerator of $\frac{5}{7}$ is bigger, so $\frac{5}{7} > \frac{3}{7}$.

2. $\mathbf{\frac{3}{10}}$**:** The numerators of the two fractions are the same, but the denominator of $\frac{3}{10}$ is smaller, so $\frac{3}{10} > \frac{3}{13}$.

3. $\mathbf{\frac{5}{4}}$**:** $\frac{1}{2} + \frac{3}{4} = \frac{1}{2} \times \frac{2}{2} + \frac{3}{4} = \frac{2}{4} + \frac{3}{4} = \frac{2+3}{4} = \frac{5}{4}$

4. $\mathbf{\frac{7}{24}}$**:** $\frac{2}{3} - \frac{3}{8} = \frac{2}{3} \times \frac{8}{8} - \frac{3}{8} \times \frac{3}{3} = \frac{16}{24} - \frac{9}{24} = \frac{16-9}{24} = \frac{7}{24}$

5. **11:** $\frac{x}{5} + \frac{2}{5} = \frac{13}{5}$

$$\frac{x}{5} + \frac{13}{5} - \frac{2}{5}$$

$$\frac{x}{5} = \frac{11}{5}$$

$$x = 11$$

6. **4:** $\frac{x}{3} - \frac{4}{9} = \frac{8}{9}$

$$\frac{x}{3} = \frac{8}{9} + \frac{4}{9}$$

$$\frac{x}{3} = \frac{12}{9}$$

$$\frac{x}{3} \times \frac{3}{3} = \frac{12}{9}$$

$$\frac{3x}{9} = \frac{12}{9}$$

$$3x = 12$$

$$x = 4$$

7. $\mathbf{\frac{5}{8}}$**:** $\frac{25}{40} = \frac{5 \times 5}{8 \times 5} = \frac{5 \times \cancel{5}}{8 \times \cancel{5}} = \frac{5}{8}$

8. $\mathbf{\frac{2}{3}}$**:** $\frac{16}{24} = \frac{2 \times 8}{3 \times 8} = \frac{2 \times \cancel{8}}{3 \times \cancel{8}} = \frac{2}{3}$

9. $\mathbf{\frac{9}{35}}$**:** $\frac{3}{10} \times \frac{6}{7} = \frac{3}{2 \times 5} \times \frac{3 \times 2}{7} = \frac{3}{\cancel{2} \times 5} \times \frac{3 \times \cancel{2}}{7} = \frac{3 \times 3}{5 \times 7} = \frac{9}{35}$

10. $\mathbf{\frac{1}{8}}$**:** $\frac{5}{14} \times \frac{7}{20} = \frac{5}{2 \times 7} \times \frac{7}{4 \times 5} = \frac{\cancel{5}}{2 \times \cancel{7}} \times \frac{\cancel{7}}{4 \times \cancel{5}} = \frac{1}{8}$

11. $\mathbf{\frac{11}{6}}: \frac{1}{6} \div \frac{1}{11} = \frac{1}{6} \times \frac{11}{1} = \frac{11}{6}$

12. $\mathbf{6}: \frac{8}{5} \div \frac{4}{15} = \frac{8}{5} \times \frac{15}{4} = \frac{2 \times 4}{5} \times \frac{3 \times 5}{4} = \frac{2 \times \cancel{4}}{\cancel{5}} \times \frac{3 \times \cancel{5}}{\cancel{4}} = \frac{6}{1} = 6$

13. $\mathbf{2}: \frac{3}{4}x = \frac{3}{2}$

$x = \frac{3}{2} \div \frac{3}{4} = \frac{3}{2} \times \frac{4}{3}$

$x = \frac{3 \times 2 \times 2}{2 \times 3} = \frac{\cancel{3} \times \cancel{2} \times 2}{\cancel{2} \times \cancel{3}} = \frac{2}{1}$

$x = 2$

14. $\mathbf{10}: \frac{x}{6} = \frac{5}{3}$

$3 \times x = 5 \times 6$

$3x = 30$

$x = 10$

15. $\mathbf{1\frac{5}{6}}: \frac{11}{6} = \frac{6+5}{6} = \frac{6}{6} + \frac{5}{6} = 1 + \frac{5}{6} = 1\frac{5}{6}$

16. $\mathbf{9\frac{1}{11}}: \frac{100}{11} = \frac{99+1}{11} = \frac{99}{11} + \frac{1}{11} = 9 + \frac{1}{11} = 9\frac{1}{11}$

17. $\mathbf{\frac{15}{4}}: 3\frac{3}{4} = 3 + \frac{3}{4} = \frac{3}{1} \times \frac{4}{4} + \frac{3}{4} = \frac{12}{4} + \frac{3}{4} = \frac{15}{4}$

18. $\mathbf{\frac{17}{3}}: 5\frac{2}{3} = 5 + \frac{2}{3} = \frac{5}{1} \times \frac{3}{3} + \frac{2}{3} = \frac{15}{3} + \frac{2}{3} = \frac{17}{3}$

19. $\mathbf{\frac{9}{10}}: \frac{\frac{3}{5}}{\frac{2}{3}} = \frac{\frac{3}{5} \times 15}{\frac{2}{3} \times 15} = \frac{9}{10}$. Alternatively, $\frac{\frac{3}{5}}{\frac{2}{3}} = \frac{3}{5} \times \frac{3}{2} = \frac{9}{10}$.

20. $\mathbf{\frac{20}{7}}: \frac{\frac{5}{7}}{\frac{1}{4}} = \frac{\frac{5}{7} \times 28}{\frac{1}{4} \times 28} = \frac{20}{7}$. Alternatively, $\frac{\frac{5}{7}}{\frac{1}{4}} = \frac{5}{7} \times \frac{4}{1} = \frac{20}{7}$

21. $\mathbf{\frac{1}{8}}: \frac{1}{2} \times \frac{1}{4} = \frac{1}{8}$

22. $\mathbf{2}: \frac{1}{2} \div \frac{1}{4} = \frac{1}{2} \times \frac{4}{1} = \frac{4}{2} = 2$

23. $\mathbf{\frac{1}{13}}:$ $\boxed{3 \times 4 = 12}$ $\frac{4}{13} \times \frac{1}{3}$ $\boxed{13 \times 1 = 13}$ $\longrightarrow$ $\frac{1}{3}$ is therefore greater than $\frac{4}{13}$.

24. $\mathbf{\frac{5}{9}}$**:** $\boxed{5\times13=65}$ $\frac{5}{9} \times \frac{7}{13}$ $\boxed{7\times9=63}$ $\longrightarrow \frac{7}{13}$ is therefore smaller than $\frac{5}{9}$.

25. **4:** Add the numerator and simplify: $\frac{13+7}{5}=\frac{20}{5}=4$.

26. $\mathbf{2\frac{1}{13}}$**:** Add the numerator and the denominator, then convert to a mixed number: $\frac{21+6}{7+6}=\frac{27}{13}=2\frac{1}{13}$.

27. $\mathbf{\frac{12(4a+b)}{a+b}}$**:** The only manipulation we can perform is to factor 12 out of the numerator:

$\frac{48a+12b}{a+b}=\frac{12(4a+b)}{a+b}$. No further simplification is possible.

28. $\mathbf{\frac{3}{2}}$**:** Factor a 3 out of the numerator and a 2 out of the denominator:

$$\frac{9g-6h}{6g-4h}=\frac{3(3g-2h)}{2(3g-2h)}.$$

Now we can cancel out the $3g - 2h$ term out of both the numerator <u>and</u> denominator:

$$\frac{3(3g-2h)}{2(3g-2h)}=\frac{3}{2}\times\frac{3g-2h}{3g-2h}=\frac{3}{2}\times1=\frac{3}{2}.$$

29. $\mathbf{\frac{171}{340}}$**:** $\frac{123}{250}$ is a little less than $\frac{125}{250}$, and so is less than $\frac{1}{2}$.

$\frac{171}{340}$ is a little more than $\frac{170}{340}$, and so is less than $\frac{1}{2}$.

$\frac{171}{340}$ is thus greater than $\frac{123}{250}$.

30. **50:** Approximate each term. $\frac{15}{58}\approx\frac{15}{60}\approx\frac{1}{4}, \frac{9}{19}\approx\frac{9}{18}\approx\frac{1}{2}$, and 403 is close to 400.

$\left(\frac{15}{58}\right)\left(\frac{9}{19}\right)403\approx\left(\frac{1}{4}\right)\left(\frac{1}{2}\right)400\approx50$. Note that the exact amount is approximately 49.369, so our estimation was <u>extremely</u> close.

31. **Yes:** Since we are only given fractions, we pick Smart Numbers. 10 is a good number to pick because it is the common denominator of the fractions 1/2 and 4/5. Mili's first generation uHear has a capacity of 10 gigabytes. Her second-generation uHear, then, has a capacity of 30 gigabytes.

Her first-generation uHear then has 5 gigabytes filled (1/2 × 10) and her second-generation uHear has 24 gigabytes filled (4/5 × 30). If she transferred the songs on the first uHear to the second, she would be at 29/30 capacity. There is enough room for the transfer.

32. **12 hours:** Because we are given an actual number in the problem, we are not allowed to pick numbers. We should assign a variable for what we are looking for: the number of hours John is awake. Let's call that total x.

Therefore, our equation will be $\frac{1}{3}x + \frac{1}{5}x + \frac{3}{10}x + 2 = x$

The common denominator of all the fractions is 30. We can multiply the equation by 30 to eliminate all the fractions:

$$30\left(\frac{1}{3}x + \frac{1}{5}x + \frac{3}{10}x + 2\right) = (x)30$$

$$10x + 6x + 9x + 60 = 30x$$

$$25x + 60 = 30x$$

$$60 = 5x$$

$$12 = x$$

John is awake for 12 total hours.

Problem Set

For problems #1–5, decide whether the given operation will yield an INCREASE, a DECREASE, or a result that will STAY THE SAME.

1. Multiply the numerator of a positive, proper fraction by $\frac{3}{2}$.

2. Add 1 to the numerator of a positive, proper fraction and subtract 1 from its denominator.

3. Multiply both the numerator and denominator of a positive, proper fraction by $3\frac{1}{2}$.

4. Multiply a positive, proper fraction by $\frac{3}{8}$.

5. Divide a positive, proper fraction by $\frac{3}{13}$.

Solve problems #6–15.

6. Simplify: $\frac{10x}{5+x}$

7. Simplify: $\frac{8(3)(x)^2(3)}{6x}$

8. Simplify: $\frac{\frac{3}{5}+\frac{1}{3}}{\frac{2}{3}+\frac{2}{5}}$

9. Simplify: $\frac{12ab^3-6a^2b}{3ab}$ (given that $ab \neq 0$)

10. Put these fractions in order from least to greatest: $\frac{9}{17}, \frac{3}{16}, \frac{19}{20}, \frac{7}{15}$

11. Put these fractions in order from least to greatest: $\frac{2}{3}, \frac{3}{13}, \frac{5}{7}, \frac{2}{9}$

12. Lisa spends $\frac{3}{8}$ of her monthly paycheck on rent and $\frac{5}{12}$ on food. Her roommate, Carrie, who earns twice as much as Lisa, spends $\frac{1}{4}$ of her monthly paycheck on rent and $\frac{1}{2}$ on food. If the two women decide to donate the remainder of their money to charity each month, what fraction of their combined monthly income will they donate?

13. Rob spends $\frac{1}{2}$ of his monthly paycheck, after taxes, on rent. He spends $\frac{1}{3}$ on food and $\frac{1}{8}$ on entertainment. If he donates the entire remainder, $500, to charity, what is Rob's monthly income, after taxes?

14. Are $\frac{\sqrt{3}}{2}$ and $\frac{2\sqrt{3}}{3}$ reciprocals?

15. Estimate to the closest integer: What is $\frac{11}{30}$ of $\frac{6}{20}$ of 120?

16.

Quantity A		Quantity B
$\frac{2}{3} \times \frac{3}{3}$	$\times$	$\frac{2}{3} \times \frac{4}{4}$

17.

Quantity A	Quantity B
$\frac{6x+6y}{3x+y}$	8

18.

An 18 oz. glass is filled with 8 oz. of orange juice. More orange juice is added so that the glass is 5 / 6 full.

Quantity A	Quantity B
Amount of orange juice added	6 oz.

1. **INCREASE:** Multiplying the numerator of a positive fraction increases the numerator. As the numerator of a positive fraction increases, its value increases.

2. **INCREASE:** As the numerator of a positive fraction increases, the value of the fraction increases. As the denominator of a positive fraction decreases, the value of the fraction also increases. Both actions will work to increase the value of the fraction.

3. **STAY THE SAME:** Multiplying or dividing the numerator and denominator of a fraction by the same number will not change the value of the fraction.

4. **DECREASE:** Multiplying a positive number by a fraction between 0 and 1 decreases the number.

5. **INCREASE:** Dividing a positive number by a fraction between 0 and 1 increases the number.

6. **CANNOT SIMPLIFY:** There is no way to simplify this fraction; it is already in simplest form. Remember, we *cannot split the denominator!*

7. **12*x*:** First, cancel terms in both the numerator and the denominator. Then combine terms.

$$\frac{8(3)(x)^2(3)}{6x} = \frac{8(\cancel{3})(x)^2(3)}{\cancel{6}2x} = \frac{\cancel{8}4(x)^2(3)}{\cancel{2}x} = \frac{4(x)^{\cancel{2}}(3)}{\cancel{x}} = 4(x)(3) = 12x$$

8. $\mathbf{\frac{7}{8}}$**:** First, add the fractions in the numerator and denominator. This results in $\frac{14}{15}$ and $\frac{16}{15}$, respectively. To save time, multiply each of the small fractions by 15, which is the common denominator of all the fractions in the problem. Because we are multiplying the numerator *and* the denominator of the whole complex fraction by 15, we are not changing its value:

$$\frac{\frac{3}{5}+\frac{1}{3}}{\frac{2}{3}+\frac{2}{5}} = \frac{\frac{9}{15}+\frac{5}{15}}{\frac{10}{15}+\frac{6}{15}} = \frac{\frac{14}{15}}{\frac{16}{15}} = \frac{\frac{14}{15}\times 15}{\frac{16}{15}\times 15} = \frac{14}{16} = \frac{7}{8}.$$

9. $\mathbf{2(2b^2 - a)}$ **or** $\mathbf{4b^2 - 2a}$**:** First, factor out common terms in the numerator. Then, cancel terms in both the numerator and denominator.

$$\frac{6ab(2b^2 - a)}{3ab} = 2(2b^2 - a) \text{ or } 4b^2 - 2a$$

10. $\mathbf{\frac{3}{16} < \frac{7}{15} < \frac{9}{17} < \frac{19}{20}}$ **:** Use Benchmark Values to compare these fractions.

$\frac{9}{17}$ is slightly more than $\frac{1}{2}$.

$\frac{19}{20}$ is slightly less than 1.

$\frac{3}{16}$ is slightly less than $\frac{1}{4}$.

$\frac{7}{15}$ is slightly less than $\frac{1}{2}$.

This makes it easy to order the fractions: $\frac{3}{16} < \frac{7}{15} < \frac{9}{17} < \frac{19}{20}$.

11. $\mathbf{\frac{2}{9} < \frac{3}{13} < \frac{2}{3} < \frac{5}{7}}$**:** Using Benchmark Values, you should notice that $\frac{3}{13}$ and $\frac{2}{9}$ are both less than $\frac{1}{2}$. $\frac{2}{3}$ and $\frac{5}{7}$ are both more than $\frac{1}{2}$. Use cross–multiplication to compare each pair of fractions:

$3 \times 9 = 27$	$\frac{3}{13} \times \frac{2}{9}$	$2 \times 13 = 26$	Thus, $\frac{3}{13} > \frac{2}{9}$.
$2 \times 7 = 14$	$\frac{2}{3} \times \frac{5}{7}$	$5 \times 3 = 15$	Thus, $\frac{2}{3} < \frac{5}{7}$.

This makes it easy to order the fractions: $\frac{2}{9} < \frac{3}{13} < \frac{2}{3} < \frac{5}{7}$.

12. $\mathbf{\frac{17}{72}}$**:** Use Smart Numbers to solve this problem. Since the denominators in the problem are 8, 12, 4, and 2, assign Lisa a monthly paycheck of \$24, since 24 us the least common multiple of the denominators. Assign her roommate, who earns twice as much, a monthly paycheck of \$48. The two women's monthly expenses break down as follows:

	Rent	Food	Remaining
Lisa	$\frac{3}{8}$ of $24 = 9$	$\frac{5}{12}$ of $24 = 10$	$24 - (9 + 10) = 5$
Carrie	$\frac{1}{4}$ of $48 = 12$	$\frac{1}{2}$ of $48 = 24$	$48 - (12 + 24) = 12$

The women will donate a total of \$17, out of their combined monthly income of \$72.

13. **\$12,000:** You cannot use Smart Numbers in this problem, because an amount is specified. This means that the total is a certain number that we are being asked to find.

First, use addition to find the fraction of Rob's money that he spends on rent, food, and entertainment:
$\frac{1}{2} + \frac{1}{3} + \frac{1}{8} = \frac{12}{24} + \frac{8}{24} + \frac{3}{24} = \frac{23}{24}$. Therefore, the \$500 that he donates to charity represents $1 - \frac{23}{24} = \frac{24-23}{24} = \frac{1}{24}$ of his total monthly paycheck. We can set up a proportion: $\frac{500}{x} = \frac{1}{24}$. Thus, Rob's monthly income is $\$500 \times 24$, or \$12,000.

14. **YES:** The product of a number and its reciprocal must equal 1. To test whether two numbers are reciprocals, multiply them. If the product is 1, they are reciprocals; if it is not, they are not:

$$\frac{\sqrt{3}}{2} \times \frac{2\sqrt{3}}{3} = \frac{2(\sqrt{3})^2}{2(3)} = \frac{6}{6} = 1$$

The numbers are thus indeed reciprocals.

15. **Approximately 13:** Use Benchmark Values to estimate: $\frac{11}{30}$ is slightly more than $\frac{1}{3}$. $\frac{6}{20}$ is slightly less than $\frac{1}{3}$. Therefore, $\frac{11}{30}$ of $\frac{6}{20}$ of 120 should be approximately $\frac{1}{3}$ of $\frac{1}{3}$ of 120, or $\frac{120}{9}$, which is slightly more than 13. (A third of 120 is 40; a third of 40 is a little over 13.)

Another technique to solve this problem would be to write the product and cancel common factors:

$$\frac{11}{30}\times\frac{6}{20}\times 120=\frac{(11)(6)(120)}{(30)(20)}=\frac{(11)(\cancel{6})(120)}{(\cancel{30}5)(20)}=\frac{(11)(\cancel{120}6)}{(5)(\cancel{20})}=\frac{66}{5}=13.2$$

Note that for estimation problems, there is no "correct" answer. The key is to arrive at an estimate that is close to the exact answer—and to do so quickly!

16. **C:** 3/3 and 4/4 are both equal to 1. Each quantity can be rewritten as 2/3 × 1, which leaves you with 2/3.

Quantity A	**Quantity B**
$\frac{2}{3}\times\frac{3}{3}=$	$\frac{2}{3}\times\frac{4}{4}=$
$\frac{2}{3}\times 1=$	$\frac{2}{3}\times 1=$
$\mathbf{\frac{2}{3}}$	$\mathbf{\frac{2}{3}}$

Therefore, **the quantities are equal.**

17. **D:** When we add fractions, you cannot split the denominator. Therefore the most that we can simplify the expression in Quantity A is $\frac{6(x+y)}{3x+y}$. But that isn't enough to tell us whether the value of this expression is more or less than 8.

For example, if $x = 2$ and $y = 1$, then Quantity A = $\frac{6(2+1)}{3(2)+1}=\frac{18}{7}$, which is less than 8. If, however, $x = 1$ and $y = -8$, then Quantity A = $\frac{6(1+(-8))}{3(1)+(-8)}=\frac{6(-7)}{3-8}=\frac{-42}{-5}=8.4$, which is greater than 8.

Therefore, **we cannot determine which quantity is greater.**

18. **A:** The easiest way to solve this problem is to find out how much liquid is in the glass *after* the orange juice is added. The glass is 5/6 full, and $5/6 \times 18 = 15$. There are 15 ounces of orange juice. There were 8 ounces of orange juice, so 7 ounces were added.

An 18 oz. glass is filled with 8 oz. of orange juice. More orange juice is added so the glass is 5/6 full.

Quantity A	Quantity B
Amount of orange juice added = **7 oz.**	6 oz.

Therefore, **Quantity A is greater.**

g

Chapter 3

of

FRACTIONS, DECIMALS, & PERCENTS

DIGITS & DECIMALS

In This Chapter . . .

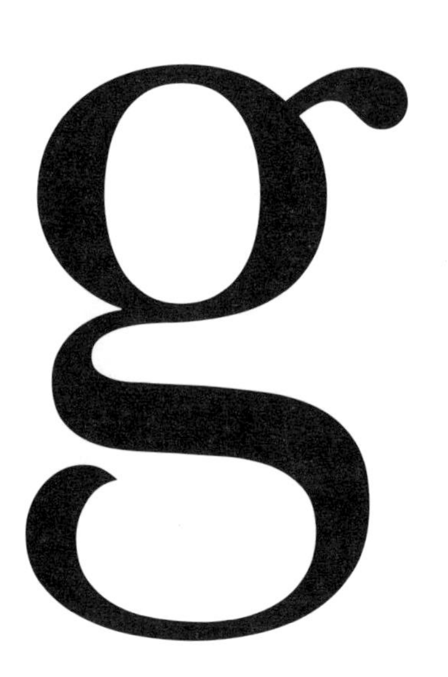

- Place Value
- Adding Zeroes to Decimals
- Powers of 10: Shifting the Decimal
- The Heavy Division Shortcut
- Decimal Operations
- Terminating vs. Non-Terminating Decimals
- Units Digit Problems

DECIMALS

GRE math goes beyond an understanding of the properties of integers (which include the counting numbers, such as 1, 2, 3, their negative counterparts, such as −1, −2, −3, and 0). The GRE also tests your ability to understand the numbers that fall in between the integers. Such numbers can be expressed as decimals. For example, the decimal 6.3 falls between the integers 6 and 7.

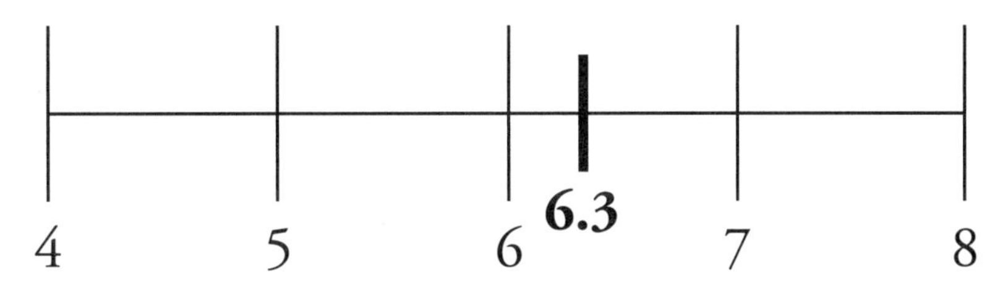

Some other examples of decimals include:

Decimals less than −1:	−3.65, −12.01, −145.9
Decimals between −1 and 0:	−0.65, −0.8912, −0.076
Decimals between 0 and 1:	0.65, 0.8912, 0.076
Decimals greater than 1:	3.65, 12.01, 145.9

Note that an integer can be expressed as a decimal by adding a decimal point and the digit 0. For example:

$8 = 8.0$ $\quad -123 = -123.0$ $\quad 400 = 400.0$

DIGITS

Every number is composed of digits. There are only ten digits in our number system:
0, 1, 2, 3, 4, 5, 6, 7, 8, 9. The term digit refers to one building block of a number; it does not refer to a number itself. For example: 356 is a number composed of three digits: 3, 5, and 6.

Integers can be classified by the number of digits they contain. For example:

2, 7, and −8 are each single-digit numbers (they are each composed of one digit).
43, 63, and −14 are each double-digit numbers (composed of two digits).
500,000 and −468,024 are each six-digit numbers (composed of six digits).
789,526,622 is a nine-digit number (composed of nine digits).

Non-integers are not generally classified by the number of digits they contain, since you can always add any number of zeroes at the end, on the right side of the decimal point:

$9.1 = 9.10 = 9.100$

That said, non-integers can be classified by how many non-zero digits it has to the right of the decimal point. For example, 0.23, 8.014, and 0.0000079 all have two non-zero digits to the right of the decimal point. (We will later discuss decimals that do not terminate, i.e., have an infinite number of non-zero digits to the right of the decimal point.)

Place Value

Every digit in a number has a particular place value depending on its location within the number. For example, in the number 452, the digit 2 is in the ones (or "units") place, the digit 5 is in the tens place, and the digit 4 is in the hundreds place. The name of each location corresponds to the "value" of that place. Thus:

2 is worth two "units" (two "ones"), or 2 ($= 2 \times 1$).
5 is worth five tens, or 50 ($= 5 \times 10$).
4 is worth four hundreds, or 400 ($= 4 \times 100$).

6	9	2	5	6	7	8	9	1	0	2	3	.	8	3	4	7
HUNDRED BILLIONS	TEN BILLIONS	ONE BILLIONS	HUNDRED MILLIONS	TEN MILLIONS	ONE MILLIONS	HUNDRED THOUSANDS	TEN THOUSANDS	THOUSANDS	HUNDREDS	TENS	UNITS OR ONES		TENTHS	HUNDREDTHS	THOUSANDTHS	TEN THOUSANDTHS

We can now write the number 452 as the **sum** of these products:

$452 = (4 \times 100) + (5 \times 10) + (2 \times 1)$
("four hundreds plus five tens plus two ones")

The chart to the left analyzes the place value of all the digits in the number:

692,567,891,023.8347

Notice that the place values to the left of the decimal all end in "-s," while the place values to the right of the decimal all end in "-ths." This is because the suffix "-ths" gives these places (to the right of the decimal) a fractional value.

Let us analyze the end of the preceding number: **0.8347**

8 is in the tenths place, giving it a value of 8 tenths, or $\frac{8}{10}$.

3 is in the hundredths place, giving it a value of 3 hundredths, or $\frac{3}{100}$.

4 is in the thousandths place, giving it a value of 4 thousandths, or $\frac{4}{1,000}$.

7 is in the ten thousandths place, giving it a value of 7 ten thousandths, or $\frac{7}{10,000}$.

To use a concrete example, 0.8 might mean eight tenths of one dollar, which would be 8 dimes or 80 cents. Additionally, 0.03 might mean three hundredths of one dollar, which would be 3 pennies or 3 cents.

Check Your Skills

1. How many digits are in 99,999?
2. In the number 4,472.1023, in what place value is the "1"?

Answers can be found on page 77.

Adding Zeroes to Decimals

Adding zeroes to the end of a decimal or taking zeroes away from the end of a decimal does not change the value of the decimal. For example: $3.6 = 3.60 = 3.6000$

Be careful, however, not to add or remove any zeroes to the left of a non-zero digit in the decimal portion of a number. Doing so will change the value of the number: $7.01 \neq 7.1$, for example.

Powers of 10: Shifting the Decimal

Place values continually decrease from left to right by powers of 10. Understanding this can help you understand the following shortcuts for multiplication and division.

When we multiply any number by a positive power of ten, we move the decimal *forward (right)* the specified number of places. This makes positive numbers larger:

$3.9742 \times 10^3 = 3{,}974.2$ Move the decimal forward 3 spaces.
$89.507 \times 10 = 895.07$ Move the decimal forward 1 space.

When we divide any number by a positive power of ten, we move the decimal *backward (left)* the specified number of places. This makes positive numbers smaller:

$4{,}169.2 \div 10^2 = 41.692$ Move the decimal backward 2 spaces.
$89.507 \div 10 = 8.3708$ Move the decimal backward 1 space.

Note that if we need to add zeroes in order to shift a decimal, we should do so:

$2.57 \times 10^6 = 2{,}570{,}000$ Add 4 zeroes at the end.
$14.29 \div 10^5 = 0.0001429$ Add 3 zeroes at the beginning.

Finally, note that negative powers of ten reverse the regular process:

$6{,}782.01 \times 10^{-3} = 6.78201$ $53.0447 \div 10^{-2} = 5{,}304.47$

We can think about these processes as **trading decimal places for powers of ten**. Think about why this is. The expression 10^{-3} is equal to 0.001. As a concrete example, if we multiply 6,782.01 by 0.001, we get a much smaller number.

For instance, all of the following numbers equal 110,700.

110.7	×	10^3
11.07	×	10^4
1.107	×	10^5
0.1107	×	10^6
0.01107	×	10^7

The first number gets smaller by a factor of 10 as we move the decimal one place to the left, but the second number gets bigger by a factor of 10 to compensate.

Check Your Skills

3. $0.0652 \times 10^{-2} = ?$
4. $\dfrac{264}{10^{-6}} = ?$
5. Put these numbers in order from least to greatest:
   a. 234×10^{-2} b. 0.234×10^{2} c. 2.34×10^{4}

Answers can be found on page 77.

The Heavy Division Shortcut

Some division problems involving decimals can look rather complex. But sometimes, we only need to find an approximate solution. In these cases, we often can save yourself time by using the Heavy Division Shortcut: move the decimals in the same direction and round to whole numbers.

What is $1{,}530{,}794 \div (31.49 \times 10^{4})$ to the nearest whole number?

Step 1: Set up the division problem in fraction form: $\dfrac{1{,}530{,}794}{31.49 \times 10^{4}}$

Step 2: Rewrite the problem, eliminating powers of 10: $\dfrac{1{,}530{,}794}{314{,}900}$

Step 3: The goal is to get a single digit to the left of the decimal in the denominator. In this problem, we need to move the decimal point backward 5 spaces. We can do this to the denominator as long as we do the same thing to the numerator. (Technically, what we are doing is dividing top and bottom by the same power of 10: 100,000.)

$$\frac{1{,}530{,}794}{314{,}900} = \frac{15.30794}{3.14900}$$

Now we have the single digit 3 to the left of the decimal in the denominator.

Step 4: Focus only on the whole number parts of the numerator and denominator and solve. $\dfrac{15.\cancel{30794}}{3.\cancel{14900}} \approx \dfrac{15}{3} \approx 5.$

An approximate answer to this complex division problem is 5. If this answer is not precise enough, keep one more decimal place and do long division (eg., $153 \div 31 \approx 4.9$).

Check Your Skills

6. What is the integer closest to $\dfrac{64{,}239{,}028}{16{,}127{,}512}$?

Answers can be found on page 77.

Decimal Operations

ADDITION AND SUBTRACTION

To add or subtract decimals, make sure to line up the decimal points. Then add zeroes to make the right sides of the decimals the same length.

4.319 + 221.8

Line up the decimal points and add zeroes.

$$\begin{array}{r} 4.319 \\ +\ 221.800 \\ \hline 226.119 \end{array}$$

10 – 0.063

Line up the decimal points and add zeroes.

$$\begin{array}{r} 10.000 \\ -\ \ 0.063 \\ \hline 9.937 \end{array}$$

Addition & Subtraction: Line up the decimal points!

MULTIPLICATION

To multiply decimals, ignore the decimal point until the very end. First, multiply the numbers as you would if they were whole numbers. Then count the *total* number of digits to the right of the decimal point in the factors. The product should have the same number of digits to the right of the decimal point.

0.02 × 1.4

Multiply normally:

$$\begin{array}{r} 14 \\ \times\ 2 \\ \hline 28 \end{array}$$

There are 3 digits to the right of the decimal point in the factors (the digits 0 and 2 in the first factor and the digit 4 in the second factor). Therefore, move the decimal point 3 places to the left in the product: 28 → 0.028.

Multiplication: In the factors, count all the digits to the right of the decimal point—then put that many digits to the right of the decimal point in the product.

If the product ends with 0, count it in this process: 0.8 × 0.5 = 0.40, since 8 × 5 = 40. Thus, 0.8 × 0.5 = 0.4.

If you are multiplying a very large number and a very small number, the following trick works to simplify the calculation: move the decimals **in the opposite direction** the same number of places.

0.0003 × 40,000 = ?

Move the decimal point RIGHT four places on the 0.0003 → 3
Move the decimal point LEFT four places on the 40,000 → 4

0.0003 × 40,000 = 3 × 4 = 12

The reason this technique works is that you are multiplying and then dividing by the same power of ten. In other words, you are **trading decimal places** in one number for decimal places in another number. This is just like trading decimal places for powers of ten, as we saw earlier.

DIVISION

If there is a decimal point in the dividend (the inner number) only, you can simply bring the decimal point straight up to the answer and divide normally.

Ex. **12.42 ÷ 3** = 4.14

$$\begin{array}{r} 4.14 \\ 3\overline{)12.42} \\ \underline{12} \\ 04 \\ \underline{3} \\ 12 \end{array}$$

However, if there is a decimal point in the divisor (the outer number), you should shift the decimal point to the right in *both the divisor and the dividend* to make the *divisor* a whole number. Then, bring the decimal point up and divide. Be sure to shift the decimal in both numbers before dividing.

Ex: **12.42 ÷ 0.3** → 124.2 ÷ 3 = 41.4

```
   41.4
3)124.2
  12
   04
    3
    12
```

Move the decimal one space to the right to make 0.3 a whole number. Then, move the decimal one space in 12.42 to make it 124.2.

Division: Always divide by whole numbers!

You can always simplify division problems that involve decimals by shifting the decimal point **in the same direction** in both the divisor and the dividend, even when the division problem is expressed as a fraction:

$$\frac{0.0045}{0.09} = \frac{45}{900}$$

Move the decimal 4 spaces to the right to make both the numerator and the denominator whole numbers.

Note that this is essentially the same process as simplifying a fraction. You are simply multiplying the numerator and denominator of the fraction by a power of ten—in this case, $\frac{10^4}{10^4}$, or $\frac{10{,}000}{10{,}000}$.

Keep track of how you move the decimal point! To simplify multiplication, you can move decimals in **opposite** directions. But to simplify division, you move decimals in the **same** direction.

Equivalently, by adding zeroes, you can express the numerator and the denominator as the same units, then simplify:

$$\frac{0.0045}{0.09} = \frac{0.0045}{0.0900} = 45 \text{ ten-thousandths} \div 900 \text{ ten-thousandths} = \frac{45}{900} = \frac{45}{900} = \frac{5}{100} = 0.05$$

Check Your Skills

7. 62.8 + 4.5768 = ?
8. 7.125 − 4.309 = ?
9. 0.00018 × 600,000 = ?
10. 85.702 ÷ 0.73 = ?

Answers can be found on pages 77–78.

Terminating vs. Non-Terminating Decimals

Repeating Decimals

Dividing an integer by another integer yields a decimal that either terminates (see below) or that never ends and repeats itself.

$2 \div 9 = ?$

$2 \div 9 = 0.2222... = 0.\overline{2}$

The bar above the 2 indicates that the digit 2 repeats infinitely.

Generally, you should just do long division to determine the repeating cycle. It is worth noting the following example patterns.

$4 \div 9 = 0.4444... = 0.\overline{4}$ $23 \div 99 = 0.2323... = 0.\overline{23}$

$\frac{1}{11} = \frac{9}{99} = 0.0909... = 0.\overline{09}$ $\frac{3}{11} = \frac{27}{99} = 0.2727... = 0.\overline{27}$

If the denominator is 9, 99, 999 or another number equal to a power of 10 minus 1, then the numerator gives you the repeating digits (perhaps with leading zeroes). Again, you can always find the decimal pattern by simple long division.

Non-Repeating Decimals

Some numbers, like $\sqrt{2}$ and π, have decimals that never end and never repeat themselves. The GRE will only ask you for approximations for these decimals (e.g., $\sqrt{2} \approx 1.4$, $\sqrt{3} \approx 1.7$, and $\pi \approx 3.14$). For numbers such as these, you can often find an estimate of the decimal using your GRE onscreen calculator.

Terminating Decimals

Occasionally the GRE asks you about properties of "terminating" decimals: that is, decimals that end. You can tack on zeroes, of course, but they do not matter. Here are some examples of terminating decimals: 0.2, 0.47, and 0.375. Terminating decimals can all be written as a ratio of integers (which might be reducible):

$$\frac{\text{Some integer}}{\text{Some power of ten}}$$

$$0.2 = \frac{2}{10} = \frac{1}{5} \qquad 0.47 = \frac{47}{100} \qquad 0.375 = \frac{375}{1000} = \frac{3}{8}$$

Positive powers of ten are composed of only 2's and 5's as prime factors. This means that when you reduce this fraction, you only have prime factors of 2's and/or 5's in the denominator. *Every terminating decimal shares this characteristic*: if, after being fully reduced, the denominator has any prime factors besides 2 or 5, then its decimal will not terminate (it will repeat). If the denominator only has factors of 2 and/or 5, then the decimal will terminate.

Check Your Skills

11. Which of the following decimals terminate? Which non-terminating decimals repeat, and which do not?

a. $\frac{11}{250}$ b. $\frac{393}{7}$ c. $\frac{1,283}{741}$ d. $\frac{\sqrt{3}}{\sqrt{2}}$

Answer can be found on page 78.

Units Digit Problems

Sometimes the GRE asks you to find a units (ones) digit of a large product, or a remainder after division by 10 (these are the same thing).

What is the units digit of $(8)^2(9)^2(3)^3$?

In this problem, you can use the Last Digit Shortcut:

To find the units digit of a product or a sum of integers, *only pay attention to the units digits of the numbers you are working with*. Drop any other digits.

This shortcut works because only units digits contribute to the units digit of the product.

STEP 1: $8 \times 8 = 6\underline{4}$	Drop the tens digit and keep only the last digit: 4.
STEP 2: $9 \times 9 = 8\underline{1}$	Drop the tens digit and keep only the last digit: 1.
STEP 3: $3 \times 3 \times 3 = 2\underline{7}$	Drop the tens digit and keep only the last digit: 7.
STEP 4: $4 \times 1 \times 7 = 2\underline{\mathbf{8}}$	Multiply the last digits of each of the products.

The units digit of the final product is 8.

Check Your Skills

Calculate the units digit of the following products:

12. $4^3 \times 7^2 \times 8$
13. 13^3
14. 15^{37}

Answers can be found on pages 78–79.

Check Your Skills Answer Key:

1. **5:** There are 5 digits in 99,999. Although there are only 9s, the 9 takes up 5 digit places (ten thousands, thousands, hundreds, tens and ones).

2. **Tenths place:** In the number 4,472.1023, the 1 is in the tenths place.

3. **0.000652:** Move the decimal to the left when you multiply by 10 raised to a negative power. In this case move the decimal to the left two places.

$$0.0652 \times 10^{-2} = 0.000652$$

4. **264,000,000:** Move the decimal to the right when dividing by 10 raised to a negative power. In this case, move the decimal to the right 6 places.

$$\frac{264}{10^{-6}} = 264{,}000{,}000$$

5. **a, b, c:**

a = 2.34
b = 23.4
c = 23,400

6. **4:** With large numbers, we can effectively ignore the smaller digits.

$$\frac{64{,}239{,}028}{16{,}127{,}512} \approx \frac{64.239028}{16.127512} \approx \frac{64}{16} \approx 4$$

Note that it is not good enough to focus on just the first digits in the numerator and denominator. That would give us $\frac{6}{1}$, or 6, which is not accurate enough.

7. **67.3768:**

$$\begin{array}{r} 62.8 \\ +4.5768 \\ \hline 67.3768 \end{array}$$

8. **2.816:**

$$\begin{array}{r} 7.125 \\ -4.309 \\ \hline 2.816 \end{array}$$

9. **108:** Trade decimal places. Change 0.00018 to 18 by moving the decimal to the right 5 places. To compensate, move the decimal of 600,000 to the left 5 places, making it 6. The multiplication problem is now:

$$18 \times 6 = 108$$

10. **117.4:** Be sure to move the decimal so that you are *dividing by whole numbers*—and be sure to move the decimal the same direction in both the dividend and the divisor. 85.702 ÷ 0.73 → 8,570.2 ÷ 73.

$$\begin{array}{r} 117.4 \\ 73\overline{)8570.2} \\ \underline{73} \\ 127 \\ \underline{73} \\ 540 \\ \underline{511} \\ 292 \\ \underline{292} \\ 0 \end{array}$$

11. **Terminating: a.; Repeating: b., c.; Non-Repeating: d.:**

$\frac{11}{250}$ has a denominator with a prime factorization of $2 \times 5 \times 5 \times 5$. Since this only includes 2s and 5s, the decimal form of the fraction will terminate. To be precise, $\frac{11}{250} = 0.044$.

In $\frac{393}{7}$, 393 is not divisible by 7, and 7 is a prime (and not a 2 or a 5). Thus the decimal will repeat infinitely: $\frac{393}{7} = 56.\overline{142857}$.

In $\frac{1,283}{741}$, the prime factorization of the denominator is $741 = 3 \times 13 \times 19$. Since this includes primes other than 2 and 5 and is fully reduced, and since the numerator and denominator are both integers, the decimal will repeat infinitely (eventually!)

In $\frac{\sqrt{3}}{\sqrt{2}}$, both the numerator and denominator are what are known as <u>irrational numbers</u>. This means they are decimals which never exhibit a repeating pattern and therefore cannot be expressed as fractions with integers.

12. **8:** Focus only on the units digit of each step of the problem:

$$4^3 = 4 \times 4 \times 4 = 6\underline{\mathbf{4}}$$
$$7^2 = 7 \times 7 = 4\underline{\mathbf{9}}$$
$$8 = \underline{\mathbf{8}}$$
$$4 \times 9 = 3\underline{\mathbf{6}}$$
$$6 \times 8 = 4\underline{\mathbf{8}}$$

13. **7:** Since we are dealing with only the units digit of the product, we can ignore the tens digit of 13 (1) and focus only on 3^3: $3 \times 3 \times 3 = 2\underline{7}$.

14. **5:** For higher exponents in units digit problems, try to find a pattern as you raise the base to higher powers.

$5^1 = \underline{5}$
$5^2 = 2\underline{5}$
$5^3 = 12\underline{5}$

Notice that the units digit is always 5? This is because $5 \times 5 = 2\underline{5}$.

Therefore $15^{37} \rightarrow 5^{37} = 5 \times 5 \times 5 \times \ldots = 2\underline{5} \times 5 \times \ldots \rightarrow 2\underline{5} \times \ldots \rightarrow \underline{5}$.

Problem Set

Solve each problem, applying the concepts and rules you learned in this section.

1. If k is an integer, and if 0.02468×10^k is greater than 10,000, what is the least possible value of k?

2. Which integer values of b would give the number $2002 \div 10^{-b}$ a value between 1 and 100?

3. Estimate to the nearest 10,000: $\dfrac{4{,}509{,}982{,}344}{5.342 \times 10^4}$

4. Simplify: $(4.5 \times 2 + 6.6) \div 0.003$

5. Simplify: $(4 \times 10^{-2}) - (2.5 \times 10^{-3})$

6. What is $4{,}563{,}021 \div 10^5$, rounded to the nearest whole number?

7. Simplify: $(0.08)^2 \div 0.4$

8. Simplify: $[8 - (1.08 + 6.9)]^2$

9. Which integer values of j would give the number $-37{,}129 \times 10^j$ a value between -100 and -1?

10. Simplify: $\dfrac{0.00081}{0.09}$

11. Determine the number of non-zero digits to the right of the decimal place for the following terminating decimals:

    a. $\dfrac{631}{100}$ b. $\dfrac{13}{250}$ c. $\dfrac{35}{50}$

12. What is the units digit of $16^4 \times 27^3$?

13.

<u>Quantity A</u>	<u>Quantity B</u>
$\dfrac{573}{10^{-2}}$	0.573×10^5

14.

Quantity A	Quantity B
$\frac{603{,}789{,}420}{13.3 \times 10^7}$	5

15.

Quantity A	Quantity B
$\left(1+\frac{2}{5}\right) \times 0.25$	0.35

1. **6:** Multiplying 0.02468 by a positive power of ten will shift the decimal point to the right. Simply shift the decimal point to the right until the result is greater than 10,000. Keep track of how many times you shift the decimal point. Shifting the decimal point 5 times results in 2,468. This is still less than 10,000. Shifting one more place yields 24,680, which is greater than 10,000.

2. **{−2, −3}:** In order to give 2002 a value between 1 and 100, we must shift the decimal point to change the number to 2.002 or 20.02. This requires a shift of either two or three places to the left. Remember that, while multiplication shifts the decimal point to the right, division shifts it to the left. To shift the decimal point 2 places to the left, we would divide by 10^2. To shift it 3 places to the left, we would divide by 10^3. Therefore, the exponent $-b = \{2, 3\}$, and $b = \{-2, -3\}$.

3. **90,000:** Use the Heavy Division Shortcut to estimate:

$$\frac{4{,}509{,}982{,}344}{53{,}420} = \frac{450{,}998.2344}{5.3420} \approx \frac{450{,}000}{5} \approx 90{,}000$$

4. **5,200:** Use the order of operations, PEMDAS (Parentheses, Exponents, Multiplication & Division, Addition & Subtraction), to simplify. Remember that the numerator acts as a parentheses in a fraction.

$4.5 \times 2 = 9$

$$\frac{9+6.6}{0.003} = \frac{15.6}{0.003} = \frac{15{,}600}{3} = 5{,}200$$

5. **0.0375:** First, rewrite the numbers in standard notation by shifting the decimal point. Then, add zeroes, line up the decimal points, and subtract.

$$\begin{array}{r} 0.0400 \\ -\underline{\ 0.0025} \\ 0.0375 \end{array}$$

6. **46:** To divide by a positive power of 10, shift the decimal point to the left. This yields 45.63021. To round to the nearest whole number, look at the tenths place. The digit in the tenths place, 6, is more than five. Therefore, the number is closest to 46.

7. **0.016:** Use the order of operations, PEMDAS (Parentheses, Exponents, Multiplication & Division, Addition & Subtraction), to simplify. Shift the decimals in the numerator and denominator so that you are dividing by an integer.

$$\frac{(0.08)^2}{0.4} = \frac{0.0064}{0.4} = \frac{0.064}{4} = 0.016$$

8. **0.0004:** Use the order of operations, PEMDAS (Parentheses, Exponents, Multiplication & Division, Addition & Subtraction), to simplify.

First, add 1.08 + 6.9 by lining up the decimal points:

$$\begin{array}{r} 1.08 \\ +\underline{6.9\ \ } \\ 7.98 \end{array}$$

Then, subtract 7.98 from 8 by lining up the decimal points, adding zeroes to make the decimals the same length:

$$\begin{array}{r} 8.00 \\ \underline{-7.98} \end{array}$$

Finally, square 0.02, by multiplying 2 × 2, and then recognizing that (0.02) × (0.02) has a total of <u>four</u> digits to the right of the decimal point.
4 ⟶ 0.0004

$$\begin{array}{r} 0.02 \\ \times\ \underline{0.02} \\ 0.0004 \end{array}$$

9. **{−3, −4}:** In order to give −37,129 a value between −100 and −1, we must shift the decimal point to change the number to −37.129 or −3.7129. This requires a shift of either three or four places to the left. Remember that multiplication shifts the decimal point to the right. To shift the decimal point 3 places to the left, we would multiply by 10^{-3}. To shift it 4 places to the left, we would multiply by 10^{-4}. Therefore, the exponent $j = \{-3, -4\}$.

10. **0.009:** Shift the decimal point 2 spaces to eliminate the decimal point in the denominator.

$$\frac{0.00081}{0.91} = \frac{0.081}{9}$$

Then divide. First, drop the 3 decimal places: 81 ÷ 9 = 9. Then put the 3 decimal places back: 0.009.

11. **a = 2, b = 2, c = 1:**

$$\frac{631}{100} = 6.\underbrace{31}_{2} \quad \frac{13}{250} = 0.0\underbrace{52}_{2} \quad \frac{35}{50} = \frac{7}{10} = 0.\underbrace{7}_{1}$$

12. **8:** We can focus on the last digits <u>only</u>: $16^4 \times 27^3 \rightarrow 6^4 \times 7^3$

$6^4 \rightarrow 6^2 \times 6^2 \rightarrow 3\underline{6} \times 3\underline{6} \rightarrow 3\underline{6} \rightarrow 6$
$7^3 \rightarrow 7^2 \times 7 \rightarrow 4\underline{9} \times 7 \rightarrow 6\underline{3} \rightarrow 3$
$6 \times 3 = 1\underline{8} \rightarrow 8$

13. **C:** When we divide by 10 raised to a negative exponent, we move the decimal to the right. 573 becomes 57,300.

<u>Quantity A</u>	<u>Quantity B</u>
$\frac{573}{10^{-2}} =$ **57,300**	0.573×10^5

When we are multiplying by 10 raised to a positive exponent, move the decimal to the right. 0.573 becomes 57,300.

<u>Quantity A</u>	<u>Quantity B</u>
57,300	$0.573 \times 10^5 =$ **57,300**

Therefore **the two quantities are equal**.

14. **B:** Quantity A looks pretty intimidating at first. The trap here is to try to find an exact value for the expression in Quantity A. Let's estimate instead:

603,789,420 is about 600,000,000.

13.3×10^7 is about 133,000,000, or even better, 130,000,000.

Quantity A	Quantity B
$\frac{603{,}789{,}420}{13.3 \times 10^7} \approx \frac{\mathbf{600{,}000{,}000}}{\mathbf{130{,}000{,}000}}$	5

Now you can cross off the zeros.

Quantity A	Quantity B
$\frac{60\,\cancel{0{,}000{,}000}}{13\,\cancel{0{,}000{,}000}} = \frac{\mathbf{60}}{\mathbf{13}}$	5

Multiply both quantities by 13.

Quantity A	Quantity B
$\frac{60}{13} \times 13 = \mathbf{60}$	$5 \times 13 = \mathbf{65}$

Therefore **Quantity B is greater.**

15. **C:** Whenever we multiply fractions or decimals, we usually prefer to convert the numbers to fractions. Simplify the parentheses in Quantity A and convert 0.25 to a fraction.

Quantity A	Quantity B
$\left(1 + \frac{2}{5}\right) \times 0.25 =$	0.35
$\left(\frac{5}{5} + \frac{2}{5}\right) \times \left(\frac{1}{4}\right) =$	
$\left(\frac{7}{5}\right) \times \left(\frac{1}{4}\right) = \frac{7}{20}$	

Now compare 7/20 and 0.35. Put 0.35 into fraction form and reduce.

Quantity A	Quantity B
$\frac{7}{20}$	$0.35 = \frac{35}{100} = \frac{7}{20}$

Therefore **the quantities are the same.**

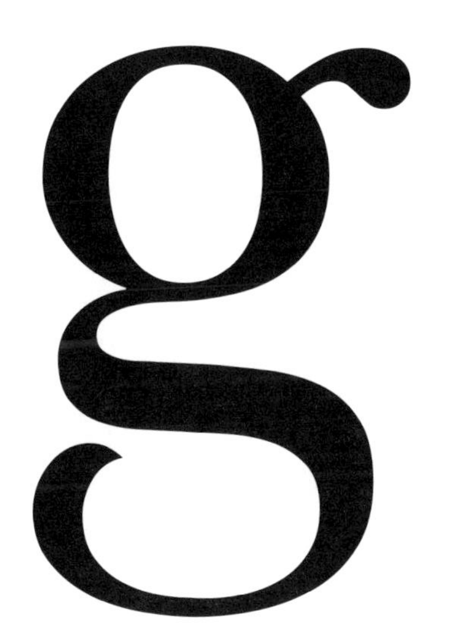

Chapter 4
of
FRACTIONS, DECIMALS, & PERCENTS

PERCENTS

In This Chapter . . .

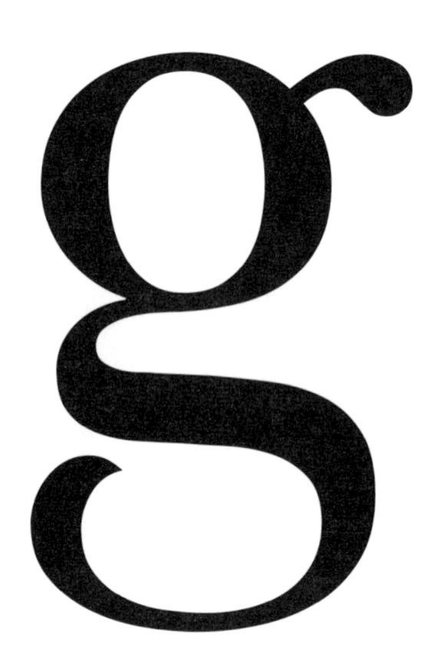

- Percents as Fractions: The Percent Table
- Benchmark Values: 10% and 5%
- Percent Increase and Decrease
- Percent Change vs. Percent of Original
- Successive Percents
- Smart Numbers: Pick 100
- Interest Formulas: Simple and Compound

PERCENTS

The other major way to express a part–whole relationship (in addition to decimals and fractions) is to use percents. Percent literally means "per one hundred." One can conceive of percents as simply a special type of fraction or decimal that involves the number 100.

75% of the students like chocolate ice cream.

This means that, out of every 100 students, 75 like chocolate ice cream. In fraction form, we write this as $\frac{75}{100}$, which simplifies to $\frac{3}{4}$.

In decimal form, we write this as 0.75 or seventy–five hundredths. Note that the last digit of the percent is in the *hundredths* place value.

One common error is to mistake 100% for 100. This is not correct. In fact, 100% means $\frac{100}{100}$, or one hundred hundredths. Therefore, 100% = 1.

Percent problems occur frequently on the GRE. The key to these percent problems frequntly is to make them concrete by picking **real numbers** with which to work.

Percents as Fractions: The Percent Table

A simple but useful way of structuring basic percent problems on the GRE is by relating percents to fractions through a percent table as shown below.

	Numbers	Percentage Fraction
PART		
WHOLE		100

A PART is some PERCENT of a WHOLE.

$$\frac{\text{PART}}{\text{WHOLE}} = \frac{\text{PERCENT}}{100}$$

Example 1: What is 30% of 80?

We are given the whole amount and the percent, and we are looking for the part. First, we fill in the percent table. Then we set up a proportion, cancel, cross–multiply, and solve:

PART	x	30
WHOLE	80	100

$$\frac{x}{80} = \frac{30\!\!\!/}{100\!\!\!/} = \frac{3}{10} \qquad 10x = 240 \qquad x = 24$$

We can also solve this problem using decimal equivalents: $(0.30)(80) = (3)(8) = 24$

Example 2: 75% of what number is 21?

We are given the part and the percent, and we are looking for the whole amount. First, we fill in the percent table. Then we set up a proportion, cancel, cross–multiply, and solve:

PART	21	75
WHOLE	x	100

$$\frac{21}{x} = \frac{75}{100} = \frac{3}{4} \qquad 3x = 84 \qquad x = 28$$

Likewise, we can also solve this problem using decimal equivalents:

$(0.75)x = 21$ then move the decimal → $75x = 2{,}100$ $x = 28$

Example 3: 90 is what percent of 40?

This time we are given the part and the whole amount, and we are looking for the percent. Note that the "part" (90) is BIGGER than the "whole" (40). That is potentially confusing but can happen, so watch the wording of the question *carefully*. Just make sure that you are taking the percent OF the "whole." Here, we are taking a percent OF 40, so 40 is the "whole."

First, we fill in the percent table. Then we set up a proportion again and solve:

PART	90	x
WHOLE	40	100

$\frac{90}{40} = \frac{9}{4} = \frac{x}{100}$ $4x = 900$ $x = 225$

90 is 225% of 40. Notice that you wind up with a percent BIGGER than 100%. That is what you should expect when the "part" is bigger than the "whole."

Check Your Skills

1. 84 is 70% of what number?
2. 30 is what percent of 50?

Answers can be found on page 97.

Benchmark Values: 10% and 5%

To find 10% of any number, just move the decimal point to the left one place.

10% of 500 is 50 10% of 34.99 = 3.499 10% of 0.978 is 0.0978

Once you know 10% of a number, it is easy to find 5% of that number: 5% is half of 10%.

10% of 300 is 30 5% of 300 is $30 \div 2 = 15$

These quick ways of calculating 10% and 5% of a number can be useful for more complicated percentages.

What is 35% of 640?

Instead of multiplying 640 by 0.35, begin by finding 10% and 5% of 640.

10% of 640 is 64 5% of 640 is $64 \div 2 = 32$

35% of a number is the same as (3 × 10% of a number) + (5% of a number).

$3 \times 64 + 32 = 192 + 32 = 224$

You can also use the benchmark values to estimate percents. For example:

Karen bought a new television, originally priced at \$690. However, she had a coupon that saved her \$67. For what percent discount was Karen's coupon?

You know that 10% of 690 would be 69. Therefore, 67 is slightly less than 10% of 690.

Check Your Skills

3. What is 10% of 145.028?
4. What is 20% of 73?

Answers can be found on page 97.

Percent Increase and Decrease

Some percent problems involve the concept of percent change. For example:

The price of a cup of coffee increased from 80 cents to 84 cents. By what percent did the price change?

Percent change problems can be solved using our handy percent table, with a small adjustment. The price *change* (84 – 80 = 4 cents) is considered the part, while the *original* price (80 cents) is considered the whole.

CHANGE	4	x
ORIGINAL	80	100

$$\frac{\text{CHANGE}}{\text{ORIGINAL}} = \frac{\text{PERCENT}}{100}$$

$\frac{4}{80} = \frac{1}{20} = \frac{x}{100}$ $\quad 20x = 100 \quad x = 5$ Thus, the price increased by 5%.

By the way, do not forget to divide by the original! The percent change is NOT 4%, which may be a wrong answer choice.

Alternatively, a question might be phrased as follows:

If the price of a \$30 shirt decreased by 20%, what was the final price of the shirt?

The whole is the original price of the shirt. The percent change is 20%. In order to find the answer, we must first find the part, which is the amount of the decrease:

CHANGE	x	20
ORIGINAL	30	100

$\frac{x}{30} = \frac{20}{100} = \frac{1}{5}$ $\quad 5x = 30 \quad x = 6$

Therefore, the price of the shirt decreased by \$6. The final price of the shirt was \$30 – \$6 = \$24.

Check Your Skills

5. A GRE score (math + verbal) increased from 1250 to 1600. By what percent did the score increase?
6. 15% of the water in a full 30 gallon drum evaporated. How much water is remaining?

Answers can be found on page 97.

Percent Change vs. Percent of Original

Looking back at the cup of coffee problem, we see that the new price (84 cents) was higher than the original price (80 cents).

We can ask what percent OF the original price is represented by the new price.

$$\frac{84}{80} = \frac{21}{20} = \frac{x}{100} \qquad 20x = 2{,}100 \qquad x = 105$$

Thus, the new price is 105% OF the original price. Remember that the percent CHANGE is 5%. That is, the new price is 5% HIGHER THAN the original price. There is a fundamental relationship between these numbers, resulting from the simple idea that the CHANGE equals the NEW value minus the ORIGINAL value, or equivalently, ORIGINAL + CHANGE = NEW:

If a quantity is increased by *x* percent, then the new quantity is (100 + *x*)% OF the original. Thus a 15% increase produces a quantity that is 115% OF the original.

We can write this relationship thus: $\text{ORIGINAL} \times \left(1 + \frac{\text{Percent Increase}}{100}\right) = \text{NEW}$

In the case of the cup of coffee, we see that $80 \times \left(1 + \frac{5}{100}\right) = 80(1.05) = 84.$

Likewise, in the shirt problem, we had a 20% decrease in the price of a $30 shirt, resulting in a new price of $24.

The new price is some percent OF the old price. Let us calculate that percent.

$$\frac{24}{30} = \frac{4}{5} = \frac{x}{100} \qquad 5x = 400 \qquad x = 80$$

Thus, the new price (20% LESS THAN the original price) is 80% OF the original price.

If a quantity is decreased by *x* percent, then the new quantity is (100 – *x*)% OF the original. Thus a 15% decrease produces a quantity that is 85% OF the original.

We can write this relationship thus: $\text{ORIGINAL} \times \left(1 - \frac{\text{Percent Decrease}}{100}\right) = \text{NEW}.$

In the case of the shirt, we see that $30 \times \left(1 - \frac{20}{100}\right) = 30(0.80) = 24.$

These formulas are all just another way of saying ORIGIANL ± CHANGE = NEW.

Example 4: What number is 50% greater than 60?

The whole amount is the original value, which is 60. The percent change (i.e., the percent "greater than") is 50%. In order to find the answer, we must first find the part, which is the amount of the increase:

CHANGE	x	50
ORIGINAL	60	100

$\frac{x}{60} = \frac{50}{100} = \frac{1}{2}$ $\quad 2x = 60 \quad x = 30$

We know that ORIGINAL ± CHANGE = NEW. Therefore, the number that is 50% greater than 60 is 60 + 30 = 90, which is also 150% of 60.

We could also solve this problem using the formula: $\text{ORIGINAL} \times \left(1 + \frac{\text{Percent Increase}}{100}\right) = \text{NEW}$

$60\left(1 + \frac{50}{100}\right) = 60(1.5) = 90$

Example 5: What number is 150% greater than 60?

The whole amount is the original value, which is 60. The percent change (i.e., the percent "greater than") is 150%. In order to find the answer, we must first find the part, which is the amount of the increase:

CHANGE	x	150
ORIGINAL	60	100

$\frac{x}{60} = \frac{150}{100} = \frac{3}{2}$ $\quad 2x = 180 \quad x = 90$

Now, x is the CHANGE, NOT the new value! **It is easy to forget to add back the original amount when the percent change is more than 100%.** Thus, the number that is 150% greater than 60 is 60 + 90 = 150, which is also 250% of 60.

We could also solve this problem using the formula: $\text{ORIGINAL} \times \left(1 + \frac{\text{Percent Increase}}{100}\right) = \text{NEW}$

$60\left(1 + \frac{150}{100}\right) = 60(2.5) = 150$

Check Your Skills

7. A plant originally cost $35. The price is increased by 20%. What is the new price?
8. 70 is 250% greater than what number?

Answers can be found on page 97–98.

Successive Percents

One of the GRE's favorite tricks involves successive percents.

If a ticket increased in price by 20%, and then increased again by 5%, by what percent did the ticket price increase in total?

Although it may seem counterintuitive, the answer is NOT 25%.

To understand why, consider a concrete example. Let us say that the ticket initially cost $100. After increasing by 20%, the ticket price went up to $120 ($20 is 20% of $100).

Here is where it gets tricky. The ticket price goes up again by 5%. However, it increases by 5% of the **NEW PRICE** of $120 (not 5% of the ***original*** $100 price). 5% of $120 is 0.05(120) = $6. Therefore, the final price of the ticket is $120 + $6 = $126, not $125.

You can now see that two successive percent increases, the first of 20% and the second of 5%, DO NOT result in a combined 25% increase. In fact, they result in a combined 26% increase (because the ticket price increased from $100 to $126).

Successive percents CANNOT simply be added together! This holds for successive increases, successive decreases, and for combinations of increases and decreases. If a ticket goes up in price by 30% and then goes down by 10%, the price has NOT in fact gone up a net of 20%. Likewise, if an index increases by 15% and then falls by 15%, it does NOT return to its original value! (Try it—you will see that the index is actually down 2.25% overall!)

A great way to solve successive percent problems is to choose real numbers and see what happens. The preceding example used the real value of $100 for the initial price of the ticket, making it easy to see exactly what happened to the ticket price with each increase. **Usually, 100 will be the easiest real number to choose for percent problems.** We will explore this in greater detail in the next section.

You could also solve by converting to decimals. Increasing a price by 20% is the same as multiplying the price by 1.20.

Increasing the new price by 5% is the same as multiplying that new price by 1.05.

Thus, you can also write the relationship this way:

$$\text{ORIGINAL} \times (1.20) \times (1.05) = \text{FINAL PRICE}$$

When you multiply 1.20 by 1.05, you get 1.26, indicating that the price increased by 26% overall.

This approach works well for problems that involve many successive steps (e.g., compound interest, which we will address later). However, in the end, it is still usually best to pick $100 for the original price and solve using concrete numbers.

Check Your Skills

9. If your stock portfolio increased by 25% and then decreased by 20%, what percent of the original value would your new stock portfolio have?

Answer can be found on page 98.

Smart Numbers: Pick 100

Sometimes, percent problems on the GRE include unspecified numerical amounts; often these unspecified amounts are described by variables.

> A shirt that initially cost d dollars was on sale for 20% off. If s represents the sale price of the shirt, d is what percentage of s?

This is an easy problem that might look confusing. To solve percent problems such as this one, simply pick 100 for the unspecified amount (just as we did when solving successive percents).

If the shirt initially cost \$100, then $d = 100$. If the shirt was on sale for 20% off, then the new price of the shirt is \$80. Thus, $s = 80$.

The question asks: d is what percentage of s, or 100 is what percentage of 80? Using a percent table, we fill in 80 as the whole amount and 100 as the part. We are looking for the percent, so we set up a proportion, cross-multiply, and solve:

PART	100	x
WHOLE	80	100

$$\frac{100}{80} = \frac{x}{100} \qquad 80x = 10{,}000 \qquad x = 125$$

Therefore, d is 125% of s.

The important point here is that, like successive percent problems and other percent problems that include unspecified amounts, this example is most easily solved by plugging in a real value. For percent problems, the easiest value to plug in is generally 100. **The fastest way to success with GRE percent problems with unspecified amounts is to pick 100 as a value.** (Note that, as we saw in the fractions chapter, if any amounts are specified, we cannot pick numbers—we must solve the problem algebraically.)

Check Your Skills

10. If your stock portfolio decreased by 25% and then increased by 20%, what percent of the original value would your new stock portfolio have?

Answer can be found on page 98.

Interest Formulas: Simple and Compound

Certain GRE percent problems require a working knowledge of basic interest formulas. The compound interest formula may look complicated, but it just expresses the idea of "successive percents" for a number of periods.

Especially for compound interest questions, be prepared to use the GRE onscreen calculator to help with the math involved!

	Formula	**Example**
SIMPLE INTEREST	Principal × Rate × Time	\$5,000 invested for 6 months at an annual rate of 7% will earn \$175 in simple interest. Principal = \$5,000, Rate = 7% or 0.07, Time = 6 months or 0.5 years. $\mathbf{Prt = \$5{,}000(0.07)(0.5) = \$175}$
COMPOUND INTEREST	$P\left(1+\frac{r}{n}\right)^{nt}$, where P = principal, r = rate (decimal) n = **number of times per year** t = **number of years**	\$5,000 invested for 1 year at a rate of 8% compounded quarterly will earn approximately \$412: $\mathbf{\$5{,}000\left(1+\frac{0.08}{4}\right)^{4(1)} = \$5{,}412}$ (or \$412 of interest)

Check Your Skills

11. Assume an auto loan in the amount of $12,000 is made. The loan carries an interest charge of 14%. What is the amount of interest owed in the first three years of the loan, assuming there is no compounding?
12. For the same loan, what is the loan balance after 3 years assuming no payments on the loan, and annual compounding?
13. For the same loan, what is the loan balance after 3 years assuming no payments, and quarterly compounding?

Answers can be found on page 98.

Check Your Skills Answer Key

1. **120:**

PART	84	70
WHOLE	x	100

$\frac{84}{x} = \frac{70}{100} = \frac{7}{10}$ $\quad 7x = 840 \quad x = 120$

2. **60:**

PART	30	x
WHOLE	50	100

$\frac{x}{100} = \frac{30}{50} = \frac{3}{5}$ $\quad 5x = 300 \quad x = 60$

3. **14.5028:** Move the decimal to the left one place. 145.028 → 14.5028

4. **14.6:** To find 20% of 73, first find 10% of 73. Move the decimal to the left one place. 73 → 7.3. 20% is twice 10%:

$7.3 \times 2 = 14.6$

5. **28%:** First find the change: 1600 – 1250 = 350.

$$\frac{\text{CHANGE}}{\text{ORIGINAL}} = \frac{350}{1250} = \frac{7}{25} = \frac{7\times4}{25\times4} = \frac{28}{100} = 28\%$$

6. **25.5:**

CHANGE	x	15
ORIGINAL	30	100

$\frac{x}{30} = \frac{15}{100} = \frac{3}{20}$ $\quad 20x = 90 \quad x = 4.5$

However, the question asks how much water is remaining. 4.5 gallons have evaporated, so 30 – 4.5 = 25.5 gallons remain.

7. **42:** Recall that $\text{ORIGIANL} \times \left(1 + \frac{\text{Percent Increase}}{100}\right) = \text{NEW}$.

$$35 \times \left(1 + \frac{20}{100}\right) = 35(1.2) = 4.2$$

8. **20:** Recall that $\text{ORIGIANAL} \times \left(1 + \frac{\text{Percent Increase}}{100}\right) = \text{NEW}$. Designate the original value x.

$$x \times \left(1+\frac{250}{100}\right)=70$$

$$3.5x=70$$

$$x=20$$

9. **100%:** Pick 100 for the original value of the portfolio. A 25% increase is:

$$100\left(1+\frac{25}{100}\right)=100(1.25)=125.$$

A 20% decrease is:

$$125\left(1-\frac{20}{100}\right)=125(0.8)=100.$$

The final value is 100. Because the starting value was also 100, the portfolio is 100% of its original value.

10. **90%:** Pick 100 for the original value of the portfolio. A 25% decrease is:

$$100\left(1-\frac{25}{100}\right)=100(0.75)=75.$$

A 20% increase is:

$$75\left(1+\frac{20}{100}\right)=75(1.2)=90.$$

The final value is 90 and the original value was 100. $\frac{90}{100}=90\%$ of the original value.

11. **$5,040:** $P \times r \times t$ = $12,000 × 14% × 3 = $5,040.

12. **$17,778.53:** $P\left(1+\frac{r}{n}\right)^{nt}$, where P – $12,000, r = 14%, n = 1 (annual compounding), and t = 3 years.

$\$12{,}000\left(1+\frac{14\%}{1}\right)^{1\times 3}=\$12{,}000\times(1.14)^3=\$17{,}778.53$ (rounded to the nearest penny). This represents $17,778.53 – $12,000 = $5,778.53 in interest.

13. **$18,132.82:** $P\left(1+\frac{r}{n}\right)^{nt}$, where P = $12,000, r = 14%, n = 4 (quarterly compounding), and t = 3 years.

$\$12{,}000\left(1+\frac{14\%}{4}\right)^{4\times 3}=\$12{,}000\times(1.035)^{12}=\$18{,}132.82$ (rounded to the nearest penny). This represents $18,132.82 – $12,000 = $6,132.82 in interest.

Problem Set

Solve the following problems. Use a percent table to organize percent problems, and pick 100 when dealing with unspecified amounts.

1. x% of y is 10. y% of 120 is 48. What is x?

2. A stereo was marked down by 30% and sold for $84. What was the pre-sale price of the stereo?

3. From 1980 to 1990, the population of Mitannia increased by 6%. From 1991 to 2000, it decreased by 3%. What was the overall percentage change in the population of Mitannia from 1980 to 2000?

4. If y is decreased by 20% and then increased by 60%, what is the new number, expressed in terms of y?

5. A 7% car loan, which is compounded annually, has an interest payment of $210 after the first year. What is the principal on the loan?

6. A bowl was half full of water. 4 cups of water were then added to the bowl, filling the bowl to 70% of its capacity. How many cups of water are now in the bowl?

7. A large tub is filled with 920 liters of alcohol and 1,800 liters of water. 40% of the water evaporates. What percent of the remaining liquid is water?

8. x is 40% of y. 50% of y is 40. 16 is what percent of x?

9. 800, increased by 50% and then decreased by 30%, yields what number?

10. If 1,600 is increased by 20%, and then reduced by y%, yielding 1,536, what is y?

11. Lori deposits $100 in a savings account at 2% interest, compounded annually. After 3 years, what is the balance on the account? (Assume Lori makes no withdrawals or deposits.)

12.

Steve uses a certain copy machine that reduces an image by 13%.

Quantity A	Quantity B
The percent of the original if Steve reduces the image by another 13%	75%

13.

y is 50% of x% of x.

Quantity A	Quantity B
y	x

14.

Quantity A	Quantity B
10% of 643.38	20% of 321.69

1. **25:** We can use two percent tables to solve this problem. Begin with the fact that y% of 120 is 48:

PART	48	y
WHOLE	120	100

$$4{,}800 = 120y$$
$$y = 40$$

Then, set up a percent table for the fact that x% of 40 is 10.

PART	10	x
WHOLE	40	100

$$1{,}000 = 40x$$
$$x = 25$$

We can also set up equations with decimal equivalents to solve:
$(0.01y)(120) = 48$, so $1.2\,y = 48$ or $y = 40$. Therefore, since we know that $(0.01x)(y) = 10$, we have:

$(0.01x)(40) = 10$ $\quad$ $40x = 1{,}000$ $\quad$ $x = 25$

2. **$120:** We can use a percent table to solve this problem. Remember that the stereo was marked down 30% from the original, so we have to solve for the original price.

CHANGE	x	30
ORIGINAL	$84 + x	100

$$\frac{x}{84+x} = \frac{30}{100} \qquad 100x = 30(84+x) \qquad 100x = 30(84) + 30x$$
$$70x = 30(84) \qquad x = 36$$

Therefore, the original price was (84 + 36) = $120.

We could also solve this problem using the formula: $\text{ORIGIANAL} \times \left(1 - \frac{\text{Percent Decrease}}{100}\right) = \text{NEW}$

$$x\left(1 - \frac{30}{100}\right) = 84 \qquad 0.7x = 84 \qquad x = 120$$

3. **2.82% increase:** For percent problems, the Smart Number is 100. Therefore, assume that the population of Mitannia in 1980 was 100. Then, apply the successive percents procedure to find the overall percent change:

From 1980–1990, there was a 6% increase: $\quad 100(1 + 0.06) = 100(1.06) = 106$
From 1991–2000, there was a 3% decrease: $\quad 106(1 - 0.03) = 106(0.97) = 102.82$
Overall, the population increased from 100 to 102.82, representing a 2.82% increase.

4. **1.28*y*:** For percent problems, the Smart Number is 100. Therefore, assign y a value of 100. Then, apply the successive percents procedure to find the overall percentage change:

(1) y is decreased by 20%: $\quad 100(1 - 0.20) = 100(0.8) = 80$
(2) Then, it is increased by 60%: $\quad 80(1 + 0.60) = 80(1.6) = 128$
Overall, there was a 28% increase. If the original value of y is 100, the new value is $1.28y$.

5. **$3,000:** We can use a percent table to solve this problem, which helps us find the decimal equivalent equation.

PART	210	7
WHOLE	x	100

$$21{,}000 = 7x$$

$$x = 3{,}000$$

6. **14:** For some problems we cannot use Smart Numbers, since the total amount can be calculated. This is one of those problems. Instead, use a percent table:

PART	$0.5x + 4$	70
WHOLE	x	100

$$\frac{0.5x+4}{x} = \frac{70}{100} = \frac{7}{10}$$

$$5x + 40 = 7x$$

$$40 = 2x$$

$$x = 20$$

The capacity of the bowl is 20 cups. There are 14 cups in the bowl {70% of 20, or 0.5(20) + 4}.

PART	4	20
WHOLE	x	100

Alternately, the 4 cups added to the bowl represent 20% of the total capacity. Use a percent table to solve for x, the whole. Since $x = 20$, there are 14 (50% of 20 + 4) cups in the bowl.

7. **54%:** For this liquid mixture problem, set up a table with two columns: one for the original mixture and one for the mixture after the water evaporates from the tub.

	Original	After Evaporation
Alcohol	920	920
Water	1,800	0.60(1,800) = 1,080
TOTAL	2,720	2,000

The remaining liquid in the tub is $\frac{1{,}080}{2{,}000}$, or 54%, water.

We could also solve for the new amount of water using the formula:

$$\text{ORIGIANAL} \times \left(1 - \frac{\text{Percent Decrease}}{100}\right) = \text{NEW}$$

$1{,}800\left(1 - \frac{40}{100}\right) = (1{,}800)(0.6) = 1{,}080$ units of water. Water is $\frac{1{,}080}{920 + 1{,}080} = \frac{1{,}080}{2{,}000} = 54\%$ of the total.

8. **50%:** Use two percent tables to solve this problem. Begin with the fact that 50% of y is 40:

PART	40	50
WHOLE	y	100

$$4{,}000 = 50y$$
$$y = 80$$

Then, set up a percent table for the fact that x is 40% of y.

PART	x	40
WHOLE	80	100

$$3{,}200 = 100x$$
$$x = 32$$

Finally, 16 is 50% of 32. We could alternatively set up equations with decimal equivalents to solve: $x = (0.4)y$ We also know that $(0.5)y = 40$, so $y = 80$ and $x = (0.4)(80) = 32$. Therefore, 16 is half, or 50%, of x.

9. **840:** Apply the successive percents procedure to find the overall percentage change:
(1) 800 is increased by 50%: $800 \times 1.5 = 1{,}200$
(2) Then, the result is decreased by 30%: $1{,}200 \times 0.7 = 840$

10. **20:** Apply the percents in succession with two percent tables.

PART	x	120
WHOLE	1,600	100

$$192{,}000 = 100x$$
$$x = 1{,}920$$

Then, fill in the "change" for the part (1,920 – 1,536 = 384) and the original for the whole (1,920).

PART	384	y
WHOLE	1,920	100

$$1{,}920y = 38{,}400$$
$$y = 20$$

Alternatively we could solve for the new number using formulas. Because this is a successive percents problem, we need to "chain" the formula: once to reflect the initial increase in the number, then twice to reflect the subsequent decrease:

$$\text{ORIGINAL} \times \left(1 + \frac{\text{Percent Increase}}{100}\right) \times \left(1 - \frac{\text{Percent Decrease}}{100}\right) = \text{NEW}$$

$$1{,}600 \times \left(1 + \frac{20}{100}\right) \times \left(1 - \frac{y}{100}\right) = 1{,}536 \qquad 1{,}920 \times \left(1 - \frac{y}{100}\right) = 1{,}536 \qquad 1{,}920 - \frac{1{,}920y}{100} = 1{,}536$$

$$1{,}920 - 1{,}536 = 19.2y \qquad 384 = 19.2y \qquad 20 = y$$

11. **$106.12:** Interest compounded annually is just a series of successive percents:
(1) 100.00 is increased by 2%: $100(1.02) = 102$
(2) 102.00 is increased by 2%: $102(1.02) = 104.04$
(3) 104.04 is increased by 2%: $104.04(1.02) \cong 106.12$

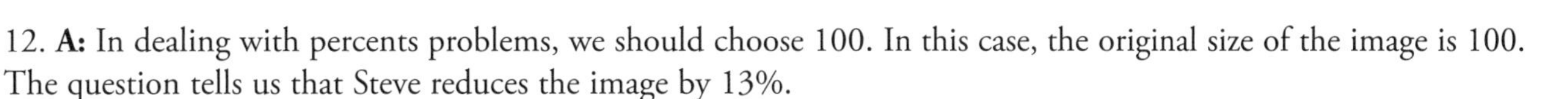

12. **A:** In dealing with percents problems, we should choose 100. In this case, the original size of the image is 100. The question tells us that Steve reduces the image by 13%.

$$100 - 0.13(100) = 100 - 13 = 87$$

So our image is at 87 percent of its original size. Quantity A tells us that we have to reduce the image size by another 13%.

If the image size is reduced by 13%, then 87% of the image remains. Multiply 87 (the current size of the image) by 0.87 (87% expressed as a decimal).

$$87 \times 0.87 = 75.69$$

Quantity A	Quantity B
The percent of the original if Steve reduces the image by another 13% = **75.69%**	75%

Therefore **Quantity A is larger.**

13. **D:** First translate the statement in the question stem into an equation.

$$y = 50\% \times \frac{x}{100} \times x \longrightarrow y = 0.5 \times \frac{x}{100} \times x = \frac{x^2}{200} \longrightarrow 200y = x^2$$

Now try to pick some easy numbers. If $y = 1$, then $x = \sqrt{200}$, which is definitely bigger than 1.

Quantity A	Quantity B
$y = \mathbf{1}$	$x = \sqrt{\mathbf{200}}$

However, if $y = 200$, then x must also equal 200.

Quantity A	Quantity B
$y = \mathbf{200}$	$x = \mathbf{200}$

y can be less than x, but y can also be equal to x. We could also choose values for which y is greater than x. Therefore **we do not have enough information** to answer the question.

14. **C:** To calculate 10% of 643.38, move the decimal to the left one place. 643.38 → 64.338

Quantity A	Quantity B
10% of 643.38 = **64.338**	20% of 321.69

To calculate 20% of 321.69, don't multiply by 0.2. Instead, find 10% first by moving the decimal to the left one place. 321.69 → 32.169

Now multiply by 2: 32.169 × 2 = 64.338

Quantity A	Quantity B
64.338	20% of 321.69 = **64.338**

Therefore **the two quantities are equal.**

Chapter 5

of

FRACTIONS, DECIMALS, & PERCENTS

FDP CONNECTIONS

In This Chapter . . .

- Converting Among Fractions, Decimals, & Percents (FDPs)
- Common FDP Equivalents
- When To Use Which Form
- FDPs and Algebraic Translations
- FDP Word Problems

FDP CONNECTIONS

Fractions, decimals, and percents are three different ways of representing the same thing: "parts of a whole."

Consider the following:

1/3 of the orange *2.5 times the distance* *110% of the sales*

In each of these instances, we're using a fraction, a decimal, or a percent to indicate that we have some portion of a whole. In fact, these different forms are very closely related. For instance, we might say that a container is 1/2 full, which is the same thing as saying that it is 50% full, or filled to 0.5 of its capacity. To illustrate, see the table below. Each row consists of a fraction, a decimal, and a percent representing the same part of a whole:

Fraction	Decimal	Percent
$\frac{1}{4}$ or $^{1}/_{4}$ or 1/4	0.25	25%
The container is 1/2 full.	The container is filled to 0.5 of its capacity.	The container is filled to 50% of its capacity.
3/2	1.5	150%

Thus, one helpful feature of fractions, decimals, and percents is that we can use whichever form is most convenient to solve a particular problem. We've already discussed fractions, decimals and percents individually. This chapter is devoted to changing from one form to another so that you can choose the form best suited to answering the question at hand.

Converting Among Fractions, Decimals, & Percents (FDPs)

From Percent to Decimal or Fraction

Percent to Decimal

As we discussed earlier, to convert from a percent to a decimal, simply move the decimal point two spots to the left.

53% becomes 0.53.

40.57% becomes 0.4057.

3% becomes 0.03.

Percent to Fraction

To convert from a percent to a fraction, remember that *per cent* literally means "per hundred," so put the percent figure over one hundred and then simplify.

45% becomes 45/100 = 9/20.

8% becomes 8/100 = 2/25.

Check Your Skills

1. Change 87% to a decimal.
2. Change 30% to a fraction.

Answers can be found on page 117.

From Decimal to Percent or Fraction

Decimal to Percent

To convert from a decimal to a percent, simply move the decimal point two spots to the right and add a percent symbol:

0.53 becomes 53%.

0.4057 becomes 40.57%.

0.03 becomes 3%.

Decimal to Fraction

To convert from decimal to fraction, it helps to remember the proper names for the digits—the first digit to the right of the decimal point is the tenths digit, next is the hundredth digit, next is the thousandth digit, and so on.

4	5	7		1	2	3	5
Hundreds	Tens	Units	•	Tenths	Hundredths	Thousandths	Ten Thousandths

The number of zeroes in the denominator should match the number of digits in the decimal (not including a possible 0 in front of the decimal point). For example:

0.3 is three tenths, or 3/10.

0.23 is twenty-three hundredths, or 23/100.

0.007 is seven thousandths, or 7/1,000.

Check Your Skills

3. Change 0.37 to a percent.
4. Change 0.25 to a fraction.

Answers can be found on page 117.

From Fraction to Decimal or Percent

Fraction to Decimal

To convert from a fraction to a decimal, long-divide the numerator by the denominator:

3/8 is $3 \div 8 = 0.375$ $\quad 8\overline{)3.000}$ = 0.375

1/4 is $1 \div 4 = 0.25$ $\quad 4\overline{)1.00}$ = 0.25

Fraction to Percent

To convert from a fraction to a percent, first convert from fraction to decimal, and then convert that decimal to a percent.

Step 1: $1/2 = 1 \div 2 = 0.50$

Step 2: $0.50 = 50\%$

Dividing the numerator by the denominator can be cumbersome and time consuming. Ideally, you should have the most basic conversions memorized before test day. A list of common FDP conversions that you should memorize appears later in the chapter.

The following chart reviews the ways to convert from fractions to decimals, from decimals to fractions, from fractions to percents, from percents to fractions, from decimals to percents, and from percents to decimals. You should practice so that each becomes natural to you.

TO → / FROM ↓	FRACTION $\frac{3}{8}$	DECIMAL 0.375	PERCENT 37.5%
FRACTION $\frac{3}{8}$		Divide the numerator by the denominator: $3 \div 8 = 0.375$ Use long division if necessary.	Divide the numerator by the denominator and move the decimal two places to the right, adding a percent symbol: $3 \div 8 = 0.375 \rightarrow 37.5\%$
DECIMAL 0.375	Use the place value of the last digit in the decimal as the denominator, and put the decimal's digits in the numerator. Then simplify: $\frac{375}{1000} = \frac{3}{8}$		Move the decimal point two places to the right and add a percent symbol: $0.375 \rightarrow 37.5\%$
PERCENT 37.5%	Use the digits of the percent for the numerator and 100 for the denominator. Then simplify: $\frac{37.5}{100} = \frac{375}{1,000} = \frac{3}{8}$	Move the decimal point two places to the left: $37.5\% \rightarrow 0.375$	

Check Your Skills

5. Change 3/5 to a decimal.
6. Change 3/8 to a percent.

Answers can be found on page 117.

Common FDP Equivalents

You should memorize the following common equivalents:

Fraction	Decimal	Percent
$1/100$	0.01	1%
$1/50$	0.02	2%
$1/25$	0.04	4%
$1/20$	0.05	5%
$1/10$	0.10	10%
$1/9$	$0.1\overline{1} \approx 0.111$	$\approx 11.1\%$
$1/8$	0.125	12.5%
$1/6$	$0.1\overline{6} \approx 0.167$	$\approx 16.7\%$
$1/5$	0.2	20%
$1/4$	0.25	25%
$3/10$	0.3	30%
$1/3$	$0.\overline{3} \approx 0.333$	$\approx 33.3\%$
$3/8$	0.375	37.5%
$2/5$	0.4	40%
$1/2$	0.5	50%

Fraction	Decimal	Percent
$3/5$	0.6	60%
$5/8$	0.625	62.5%
$2/3$	$0.\overline{6} \approx 0.667$	$\approx 66.7\%$
$7/10$	0.7	70%
$3/4$	0.75	75%
$4/5$	0.8	80%
$5/6$	$0.8\overline{3} \approx 0.833$	$\approx 83.3\%$
$7/8$	0.875	87.5%
$9/10$	0.9	90%
$1/1$	1	100%
$5/4$	1.25	125%
$4/3$	$1.\overline{3} \approx 1.33$	133%
$3/2$	1.5	150%
$7/4$	1.75	175%

When To Use Which Form

Fractions are good for cancelling factors in multiplication and division. They are also the best way of exactly expressing proportions that do not have clean decimal equivalents, such as 1/7. Switch to fractions if there is a handy fractional equivalent of the decimal or percent and/or you think you can cancel a lot of factors.

What is 37.5% of 240?

If you simply convert the percent to a decimal and multiply, you will have to do a fair bit of arithmetic:

$$\begin{array}{r} 0.375 \\ \times 240 \\ \hline 0 \\ 15000 \\ 75000 \\ \hline 90.000 \end{array}$$

Alternatively, you can recognize that $0.375 = \frac{3}{8}$.

So we have $(0.375)(240) = \left(\frac{3}{\cancel{8}}\right)\left(\cancel{240}^{30}\right) = 3(30) = 90.$

This is much faster!

A dress is marked up $16\frac{2}{3}\%$ to a final price of \$140. What is the original price of the dress?

From the previous page, we know that $16\frac{2}{3}\%$ is equivalent to $\frac{1}{6}$. Thus, adding $\frac{1}{6}$ of a number to itself is the same thing as multiplying by $1 + \frac{1}{6} = \frac{7}{6}$:

$\frac{7}{6}x = 140 \qquad x = \left(\frac{6}{7}\right)140 = \left(\frac{6}{\cancel{7}}\right)\cancel{140}^{20} = 120.$ The original price is \$120.

Decimals, on the other hand, are good for estimating results or for comparing sizes. The reason is that the basis of comparison is equivalent (there is no denominator). The same holds true for **percents**. The implied denominator is always 100, so you can easily compare percents (of the same whole) to each other.

To convert certain fractions to decimals or percents, multiply the numerator and the denominator by the same number:

$$\frac{17}{25} = \frac{17 \times 4}{25 \times 4} = \frac{68}{100} = 0.68 = 68\%$$

This process is faster than long division, but it only works when the denominator has only 2s and/or 5s as factors (as we learned earlier, fractions with denominators containing prime factors other than 2s and 5s will be non-terminating, and therefore cannot be represented exactly by decimals or percents).

In some cases, you might find it easier to compare a series of fractions by giving them all a common denominator, rather than by converting them all to decimals or percents. The general rule is this: **prefer fractions for doing multiplication or division, but prefer decimals and percents for doing addition or subtraction, for estimating numbers, or for comparing numbers**.

FDPs and Algebraic Translations

Fractions, decimals, and percents show up in many Algebraic Translations problems. Make sure that you understand and can apply the very common translations below:

In the Problem:	Translation:
X percent	$\frac{X}{100}$
of	Multiply
of Z	Z is the Whole (percents)
Y is X percent of Z	Y is the Part, and Z is the Whole $Y = \left(\frac{X}{100}\right)Z$ $\text{Part} = \left(\frac{\text{Percent}}{100}\right) \times \text{Whole}$
Y is X percent of Z	Alternative: $\frac{Y}{Z} = \frac{X}{100}$ $\frac{\text{Part}}{\text{Whole}} = \frac{\text{Percent}}{100}$
A is $\frac{1}{6}$ of B	$A = \left(\frac{1}{6}\right)B$
C is 20% of D	$C = (0.20)D$
E is 10% greater than F	$E = \left(1 + \frac{10}{100}\right)F = (1.1)F$
G is 30% less than H	$G = \left(1 - \frac{30}{100}\right)H = (0.70)H$
The dress cost \$$J$. Then it was marked up 25% and sold. What is the profit?	Profit = Revenue – Cost Profit = $(1.25)J - J$ Profit = $(0.25)J$

FDP Word Problems

As we mentioned earlier, the purpose of fractions, decimals, and percents is to represent the proportions between a part and a whole.

Most FDP Word Problems hinge on these fundamental relationships:

$$\text{Part} = \text{Fraction} \times \text{Whole}$$

$$\text{Part} = \text{Decimal} \times \text{Whole}$$

$$\text{Part} = \frac{\text{Percent}}{100} \times \text{Whole}$$

In general, these problems will give you two parts of the equation and ask you to figure out the third.

Let's look at three examples:

> A quarter of the students attended the pep rally. If there are a total of 200 students, how many of them attended the pep rally?

In this case, we are told the fraction and the total number of students. We are asked to find the number of students who attended the pep rally.

$a = (1/4)(200)$
$a = 50$

Fifty students attended the pep rally.

> At a certain pet shop, there are four kittens, two turtles, eight puppies, and six snakes. If there are no other pets, what percentage of the store's animals are kittens?

Here we are told the part (there are four kittens) and the whole (there are $4 + 2 + 8 + 6 = 20$ animals total). We are asked to find the percentage.

$4 = x(20)$
$4 \div 20 = x$
$0.2 = x$
$x = 20\%$

Twenty percent of the animals are kittens.

> Sally receives a commission equal to thirty percent of her sales. If Sally earned $4,500 in commission last month, what were her total sales?

Here we are given the part, and told what percent that part is, but we don't know the whole. We are asked to solve for the whole.

$4{,}500 = 0.30s$
$4{,}500 \div 0.30 = s$
$s = 15{,}000$

Her total sales for the month were $15,000.

Tip: If in doubt—sound it out! Do you ever get confused on how exactly to set up an equation for a word problem? If so, you're not alone! For instance, consider the following problem:

x is forty percent of what number?

First, let's assign a variable to the number we're looking for—let's call it y.

Do we set this up as $40\% \times x = (y)$, or $x = 40\% \times (y)$?

If you are unsure of how to set up this equation, try this—say it aloud or to yourself. Often, that will clear up any confusion, and put you on the right track.

Let's illustrate, using our two options.

	Equation	Read out loud as...
Option 1:	$40\% \times x = (y)$	40% of $x = y$
Option 2:	$x = 40\% \ (y)$	x is 40% of y

Now, it's much easier to see that the second option, $x = 40\% \ (y)$, is the equation that represents our original question.

Check Your Skills

Write the following sentences as equations.

7. x is 60% of y.
8. 1/3 of a is b.
9. y is 25% of what number?

Answers can be found on pages 117.

Typical Complications

Now let's take those three problems and give them a typical GRE twist.

"A quarter of the students attended the pep rally. If there are a total of 200 students, how many of them did not attend the pep rally?"

Notice here that the fraction we are given, one quarter, represents the students who did attend the pep rally, but we are asked to find the number that did *not* attend the pep rally.

Here are two ways we can solve this:

1. Find the value of one quarter and subtract from the whole.

$a = (1/4)(200)$
$a = 50$

Once we figure out 50 students did attend, we can see that $200 - 50 = 150$, so 150 did not attend.

OR

2. Find the value of the remaining portion.

If 1/4 did attend, that must mean 3/4 did not attend:

$n = (3/4)(200)$
$n = 150$

"At a certain pet shop, there are four kittens, two turtles, eight puppies, and six snakes. If there are no other pets, what percentage of the store's animals are kittens or puppies?"

Here we are asked to combine two different elements. We can take either of two approaches.

1. Figure each percentage out separately and then add.

$4 = x(20)$
$0.2 = x$

$8 = y(20)$
$0.4 = y$

$0.2 + 0.4 = 0.6$

Kittens and puppies represent 60% of the animals.

OR

2. Add the quantities first and then solve.

There are four kittens and eight puppies, for a total of $4 + 8 = 12$ of these animals:

$12 = x(20)$
$0.6 = x$

Kittens and puppies represent 60% of the animals.

"Sally receives a monthly salary of $1,000 plus a 30% commission of her total sales. If Sally earned $5,500 last month, what were her total sales?"

In this case, a constant ($1,000) has been added in to the proportion equation.

Her salary = $1,000 + 0.30(total sales)

$5{,}500 = 1000 + 0.3(s)$
$4{,}500 = 0.3(s)$
$15{,}000 = s$

Alternatively, we could subtract out Sally's $1,000 salary from her earnings of $5,500 first, to arrive at the portion of her income ($5,500 – $1,000 = $4,500) derived from her 30% commission and proceed from there.

Check Your Skills

10. A water drum is filled to 1/4 of its capacity. If another 30 gallons of water were added, the drum would be filled. How many gallons of water does the drum currently contain?

Answer can be found on page 117.

Check Your Skills Answer Key

1. **0.87:** Shift the decimal two places to the left.
87% becomes 0.87.

2. **3/10:** Divide the percent figure by 100, then simplify.
30% becomes 30/100, which reduces to 3/10.

3. **37%:** Shift the decimal two places to the right and add a percent sign.
0.37 becomes 37%.

4. **1/4:** Notice that the decimal has two digits to the right of the decimal place.
0.25 is 25 hundredths, so it becomes 25/100, which reduces to 1/4.

5. **0.6:** Long-divide the numerator by the denominator.
$3/5$ is $3 \div 5 = 0.6$

6. **37.5%:** Long-divide the numerator by the denominator, shift the decimal two places to the right, and add a percent sign.
Step 1: $3/8$ is $3 \div 8 = 0.375$
Step 2: $0.375 = 37.5\%$

7. $\boldsymbol{x = 0.6y}$**:**
$x = 60\% \ (y)$
$x = 0.6y$

8. $\boldsymbol{(1/3)a = b}$**.**

9. $\boldsymbol{y = 0.25x}$**:** Let x equal the number in question:
$y = 25\% \ (x)$
$y = 0.25(x)$

10. **10 gallons:** Let x be the capacity of the water drum. If the drum is 1/4 full, and 30 gallons would make it full, then $30 = (1 - 1/4)x$, which means:

$$30 = \frac{3}{4}x$$

Divide both sides by 3/4. This is equivalent to multiplying by 4/3.

$$30 = \frac{3}{4}x$$

$$\frac{4}{3} \times 30 = x$$

$$\frac{4}{\cancel{3}} \times \cancel{30}\,10 = x$$

$$40 = x$$

If the total capacity is 40 gallons and the drum is 1/4 full, then the drum currently contains $1/4 \times 40 = 10$ gallons.

Problem Set

1. Express the following as fractions: 2.45 0.008

2. Express the following as fractions: 420% 8%

3. Express the following as decimals: $\frac{9}{2}$ $\frac{3{,}000}{10{,}000}$

4. Express the following as decimals: $1\frac{27}{4}$ $12\frac{8}{3}$

5. Express the following as percents: $\frac{1{,}000}{10}$ $\frac{25}{9}$

6. Express the following as percents: 80.4 0.0007

7. Order from least to greatest: $\frac{8}{18}$ 0.8 40%

8. Order from least to greatest: 1.19 $\frac{120}{84}$ 131.44%

9. Order from least to greatest: $2\frac{4}{7}$ 2400% 2.401

10. Order from least to greatest ($x \neq 0$): $\frac{50}{17}x^2$ $2.9x^2$ $(x^2)(3.10\%)$

11. Order from least to greatest: $\frac{500}{199}$ 248,000% 2.9002003

12. What number is 62.5% of 192?

13. 200 is 16% of what number?

For problems #14–15, express your answer in terms of the variables given (X, Y, and possibly Z).

14. What number is X percent of Y?

15. X is what percent of Y?

16. For every 1,000,000 toys sold, 337,000 are action figures.

Quantity A	Quantity B
Percent of toys sold that are action figures	33.7%

17.

Quantity A	Quantity B
$10^{-3} \times \left(\frac{0.002}{10^{-3}}\right)$	0.02

18. $1,600 worth of $20 bills are stacked up and reach 32 inches high. $1,050 worth of $10 bills are also stacked up (assume all denominations are the same height.)

Quantity A	Quantity B
The percent by which the height of the stack of $10 bills is greater than that of the stack of $20 bills	33.5%

1. To convert a decimal to a fraction, write it over the appropriate power of ten and simplify.

$$2.45 = 2\frac{45}{100} = \mathbf{2\frac{9}{20}} \text{ (mixed)} = \mathbf{\frac{49}{20}} \text{ (improper)}$$

$$0.008 = \frac{8}{1{,}000} = \mathbf{\frac{1}{125}}$$

2. To convert a percent to a fraction, write it over a denominator of 100 and simplify.

$$420\% = \frac{420}{100} = \mathbf{\frac{21}{5}} \text{ (improper)} = \mathbf{4\frac{1}{5}} \text{ (mixed)}$$

$$8\% = \frac{8}{100} = \mathbf{\frac{2}{25}}$$

3. To convert a fraction to a decimal, divide the numerator by the denominator.

$$\frac{9}{2} = 9 \div 2 = \mathbf{4.5}$$

It often helps to simplify the fraction BEFORE you divide:

$$\frac{3{,}000}{10{,}000} = \frac{3}{10} = \mathbf{0.3}$$

4. To convert a mixed number to a decimal, simplify the mixed number first, if needed.

$$1\frac{27}{4} = 1 + 6\frac{3}{4} = 7\frac{3}{4} = \mathbf{7.75}$$

$$12\frac{8}{3} = 12 + 2\frac{2}{3} = \mathbf{14.\overline{6}}$$

5. To convert a fraction to a percent, rewrite the fraction with a denominator of 100.

$$\frac{1{,}000}{10} = \frac{10{,}000}{100} = \mathbf{10{,}000\%}$$

Alternatively, we can convert the fraction to a decimal and shift the decimal point two places to the right and add a percent symbol.

$$\frac{25}{9} = 25 \div 9 = 2.7777... = 2.\overline{7} = \mathbf{277.\overline{7}}\%$$

6. To convert a decimal to a percent, shift the decimal point two places to the right and add a percent symbol.

$80.4 =$ **8,040%**
$0.0007 =$ **0.07%**

7. $\mathbf{40\% < \frac{8}{18} < 0.8}$**:** To order from least to greatest, express all the terms in the same form.

$$\frac{8}{18} = \frac{4}{9} = 0.4444... = 0.\overline{4}$$

$0.8 = 0.8$
$40\% = 0.4$
$0.4 < 0.\overline{4} < 0.8$

Alternately, we can use FDP logic and Benchmark Values to solve this problem: $\frac{8}{18}$ is $\frac{1}{18}$ less than $\frac{1}{2}$. 40% is 10% (or $\frac{1}{10}$) less than $\frac{1}{2}$. Since $\frac{8}{18}$ is a smaller piece away from $\frac{1}{2}$, it is closer to $\frac{1}{2}$ and therefore larger than 40%. 0.8 is clearly greater than $\frac{1}{2}$. Therefore, $40\% < \frac{8}{18} < 0.8$.

8. **$1.19 < 131.44\% < \frac{120}{84}$:** To order from least to greatest, express all the terms in the same form.

$1.19 = 1.19$
$\frac{120}{84} = \frac{10}{7} \approx 1.4286$
$131.44\% = 1.3144$
$1.19 < 1.3144 < 1.4286$

9. **$2.401 < 2\frac{4}{7} < 2400\%$:** To order from least to greatest, express all the terms in the same form.

$2\frac{4}{7} \approx 2.5714$
$2{,}400\% = 24$
$2.401 = 2.401$

Alternately, we can use FDP logic and Benchmark Values to solve this problem: 2400% is 24, which is clearly the largest value. Then, we can use Benchmark Values to compare $2\frac{4}{7}$ and 2.401. Since the whole number portion, 2, is the same, just compare the fraction parts. $\frac{4}{7}$ is greater than $\frac{1}{2}$. 0.401 is less than $\frac{1}{2}$. Therefore, $2\frac{4}{7}$ must be greater than 2.401. So, $2.401 < 2\frac{4}{7} < 2{,}400\%$.

10. **$(x^2)(3.10\%) < 2.9x^2 < \frac{50}{17}x^2$:** To order from least to greatest, express all the terms in the same form. (Note that, since x^2 is a positive term common to all the terms you are comparing, you can ignore its presence completely. If the common term were negative, then the order would be reversed.)

$\frac{50}{17} = 2\frac{16}{17} \approx 2.94$ (You can find the first few digits of the decimal by long division.)

$2.9 = 2.9$
$3.10\% = 0.0310$
$0.0310 < 2.9 < 2.94$

Alternately, we can use FDP logic and Benchmark Values to solve this problem: 3.10% is 0.0310, which is clearly the smallest value. Then, we compare 2.9 and $2\frac{16}{17}$ to see which one is closer to 3. 2.9 is $\frac{1}{10}$ away from 3. $2\frac{16}{17}$ is $\frac{1}{17}$ away from 3. Since $\frac{1}{17}$ is smaller than $\frac{1}{10}$, $2\frac{16}{17}$ is closest to 3; therefore, it is larger. So, $3.10\% < 2.9 < \frac{50}{17}$.

11. $\mathbf{\frac{500}{199}}$ **< 2.9002003 < 248,000%:** To order from least to greatest, express all the terms in the same form.

$\frac{500}{199} \approx 2.51$ (You can find the first few digits of the decimal by long division.)

248,000% = 2,480
2.9002003 = 2.9002003

Alternately, we can use FDP logic and Benchmark Values to solve this problem: 248,000% = 2,480, which is clearly the largest value. $\frac{500}{199}$ is approximately $\frac{500}{200}$, or $\frac{5}{2}$, which is 2.5. This is clearly less than 2.9002003. Therefore, $\frac{500}{199}$< 2.9002003 < 248,000%.

12. **120:** This is best handled as a percent-to-decimal conversion problem. If we simply recognize that $62.5\% = 0.625 = \frac{5}{8}$, this problem will become much easier: $\frac{5}{8} \times 192 = \frac{5}{1} \times 24 = 120$. Multiplying 0.625 × 240 will take much longer to complete.

13. **1,250:** This is best handled as a percent-to-decimal conversion problem. If we simply recognize that $16\% = 0.16 = \frac{16}{100} = \frac{4}{25}$, this problem will become much easier: $\frac{4}{25}x = 200$, so $x = 200 \times \frac{25}{4} = 50 \times 25 = 1,250$. Dividing out 200 ÷ 0.16 would likely take longer to complete.

14. $\mathbf{\frac{XY}{100}}$**:** We can use decimal equivalents. X percent is $\frac{X}{100}$, and we simply need to multiply by Y.

Alternatively we can set up a table and solve for the unknown (in this case, we will call it Z):

PART	Z	X
WHOLE	Y	100

$100Z = XY$

$Z = \frac{XY}{100}$

15. $\mathbf{\frac{100X}{Y}}$**:** We can use decimal equivalents. X equals some unknown percent of Y (call it Z percent), so $X = \frac{Z}{100} \times Y$, and we simply solve for Z: $\frac{100X}{Y} = Z$.

Alternatively we can set up a table and solve for the unknown Z:

PART	X	Z
WHOLE	Y	100

$100X = ZY$

$Z = \frac{100}{Y}$

16. **C:** Simplify Quantity A. We can divide the number of action figures by the total number of toys to find the percentage of action figures.

Quantity A	Quantity B
Percent of toys sold that are action figures = $\mathbf{\frac{337{,}000}{1{,}000{,}000}}$	33.7%

A percentage is defined as being out of 100, so reduce the fraction until the denominator is 100.

Quantity A	Quantity B
$\frac{337{,}000}{1{,}000{,}000} = \frac{337{,}\not{0}\not{0}\not{0}}{1{,}000{,}\not{0}\not{0}\not{0}} =$ $\frac{337}{1{,}000} = \mathbf{\frac{33.7}{100}}$	33.7%

Because the denominator is 100, the number in the numerator is the percent. So action figures are 33.7% of the total number of toys. **The two quantities are equal.**

17. **B:** Take a close look at the expression in Quantity A: 0.002 is first divided by 10^{-3}, and then multiplied by 10^{-3}. The net effect is the same as multiplying by 1. The two 10^{-3} terms cancel out.

Quantity A	Quantity B
$10^{-3}\times\left(\frac{0.002}{10^{-3}}\right) = 0.002\times\frac{10^{-3}}{10^{-3}} =$ $0.002\times 1 = \mathbf{0.002}$	**0.02**

Therefore **Quantity B is larger.**

18. **B:** Because all bills have the same height, we can compare the number of bills in each stack directly to determine the percent increase in height. The number of \$20 bills in a stack with a value of \$1,600 is:

1600/20 = 80

The number of \$10 bills in a stack with a value of \$1,050 is

1,050/10 = 105

Plug these values into the percent change formula to evaluate Quantity A.

Quantity A	Quantity B
The percent by which the height of the stack of $10 bills is greater than that of the stack of $20 bills $= \frac{105-80}{80} = \frac{25}{80} = \mathbf{\frac{5}{16}}$	33.5%

Now compare the two quantities. $\frac{5}{16} < \frac{5}{15}$, so Quantity A must be less than $\frac{1}{3}$. Recall that $\frac{1}{3}$ is $33.\overline{3}\%$ as a percent, so 33.5% is slightly larger than $\frac{1}{3}$. Therefore the value in Quantity A must be less than 33.5%.

Thus **Quantity B is greater**.

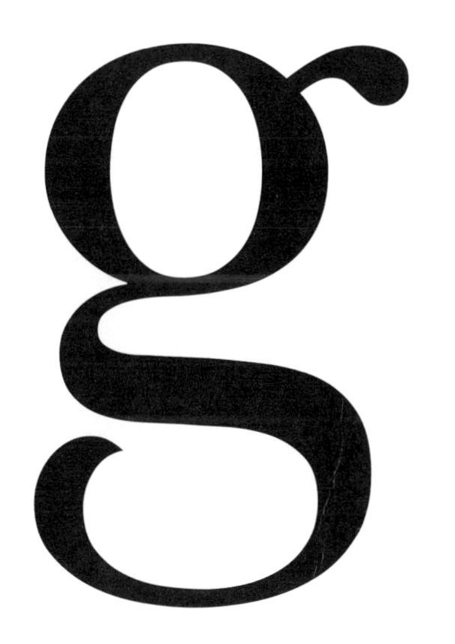

Chapter 6

of

FRACTIONS, DECIMALS, & PERCENTS

DRILL SETS

In This Chapter . . .

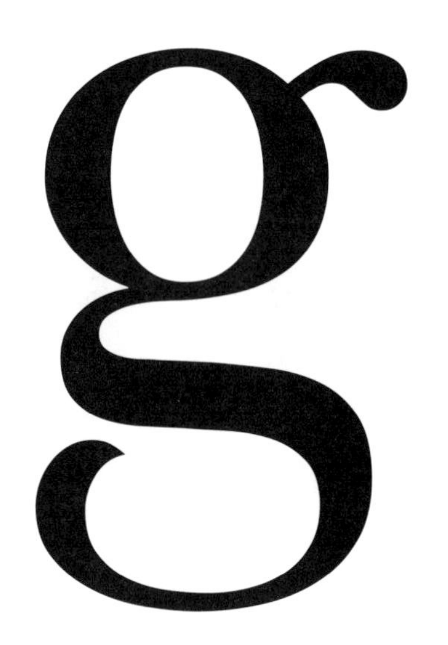

- FDPs Drill Sets

DRILL SET 1:

Drill 1: For each of the following pairs of fractions, decide which fraction is larger:

1. $\frac{1}{4}, \frac{3}{4}$

2. $\frac{1}{5}, \frac{1}{6}$

3. $\frac{53}{52}, \frac{85}{86}$

4. $\frac{7}{9}, \frac{6}{10}$

5. $\frac{700}{360}, \frac{590}{290}$

Drill 2: Add or subtract the following fractions:

1. $\frac{2}{7} + \frac{3}{7} =$

2. $\frac{5}{8} - \frac{4}{8} =$

3. $\frac{7}{9} - \frac{2}{9} =$

4. $\frac{9}{11} + \frac{20}{11} =$

5. $\frac{3}{4} - \frac{10}{4} =$

Drill 3: Add or subtract the following fractions:

1. $\frac{2}{3} + \frac{5}{9}$

2. $\frac{7}{8} - \frac{5}{4}$

3. $\frac{4}{9} + \frac{8}{11}$

4. $\frac{20}{12} - \frac{5}{3}$

5. $\frac{1}{4} + \frac{4}{5} + \frac{5}{8}$

DRILL 4: Solve for x in the following equations:

1. $\frac{1}{5} + \frac{x}{5} = \frac{4}{5}$

2. $\frac{x}{8} - \frac{3}{8} = \frac{10}{8}$

3. $\frac{x}{6} + \frac{5}{12} = \frac{11}{12}$

4. $\frac{2}{7} - \frac{x}{21} = -\frac{2}{21}$

5. $\frac{2}{5} + \frac{x}{8} = \frac{31}{40}$

DRILL SET 2:

Drill 1: Simplify the following fractions:

1. $\frac{6}{9}$

2. $\frac{12}{28}$

3. $\frac{24}{36}$

4. $\frac{35}{100}$

5. $\frac{6x}{70}$

Drill 2: Multiply or divide the following fractions (the final answer must be simplified):

1. $\frac{4}{7} \times \frac{8}{9}$

2. $\frac{9}{4} \div \frac{2}{3}$

3. $\frac{7}{15} \div \frac{8}{5}$

4. $\frac{5}{12} \times \frac{8}{10}$

5. $\frac{3x}{13} \times \frac{5}{6}$

Drill 3: Multiply or divide the following fractions (the final answer must be simplified):

1. $\frac{14}{20}\times\frac{15}{21}$
2. $\frac{6}{25}\div\frac{9}{10}$
3. $\frac{4}{21}\times\frac{14}{13}\times\frac{5}{8}$
4. $\frac{3}{11}\div\frac{3}{11}$
5. $\frac{57}{63}\times\frac{0}{18}$

Drill 4: Solve for x in the following equations:

1. $\frac{5}{3}x=\frac{3}{7}$
2. $\frac{2}{x}=\frac{7}{3}$
3. $\frac{6}{11}x=\frac{10}{33}$
4. $\frac{2x}{13}=\frac{3}{7}$
5. $\frac{3x}{4}-\frac{5}{6}=\frac{17}{12}$

DRILL SET 3:

DRILL 1: Convert the following improper fractions to mixed numbers:

1. $\frac{9}{4}$
2. $\frac{31}{7}$
3. $\frac{47}{15}$
4. $\frac{70}{20}$
5. $\frac{91}{13}$

Drill 2: Convert the following mixed numbers to improper fractions:

1. $3\,^{2}/_{3}$
2. $2\,^{1}/_{6}$
3. $6\,^{3}/_{7}$
4. $4\,^{5}/_{9}$
5. $12\,^{5}/_{12}$

DRILL SET 4:

DRILL 1: Evaluate the following expressions:

1. $6.75 \times 10^{3} =$
2. $72.12 \times 10^{-4} =$
3. $2{,}346 \times 10^{-3} =$
4. $27.048 \times 10^{2} =$
5. $54.197/10^{2} =$

DRILL 2: Evaluate the following expressions:

1. $1 + 0.2 + 0.03 + 0.004 =$
2. $0.48 + 0.02 =$
3. $1.21 + 0.38 =$
4. $-0.02 + 0.35 =$
5. $0.370 + 0.042 =$

DRILL 3: Evaluate the following expressions:

1. $0.27 \times 2 =$
2. $0.403 \times 3 =$
3. $0.2 \times 0.2 =$
4. $20 \times 0.85 =$
5. $0.04 \times 0.201 =$

DRILL 4: Evaluate the following expressions:

1. $2.1 \times 0.08 =$
2. $0.063 \times 0.40 =$
3. $0.03 \times 0.005 =$
4. $0.75\ (80) + 0.50\ (20) =$
5. $100 \times 0.01 \times 0.01 =$

DRILL 5: Evaluate the following expressions:

1. $4/0.2 =$
2. $12.6/0.3 =$
3. $3.20/0.04 =$
4. $0.49/0.07 =$
5. $6/0.5 =$

DRILL SET 5:

DRILL 1: Fill in the missing information in the chart below:

Fraction	Decimal	Percent
1/100	0.01	1%
1/20		
	0.1	
1/8		
	0.2	
		25%
	0.3	
		33.33...%
3/8		
		40%
1/2		
	0.6	
		66.66...%
		70%
	0.75	
4/5		
	0.875	
9/10		
		100%

DRILL 2:

1. Convert 45% to a decimal.
2. Convert 70% to a fraction.
3. Convert 13.25% to a decimal.
4. Convert 36% to a fraction.
5. Convert 0.02% to a decimal.

DRILL 3:

1. Convert 0.20 to a percent.
2. Convert 0.55 to a fraction.
3. Convert 0.304 to a percent.
4. Convert 0.455 to a fraction.
5. Convert 0.375 to a fraction.

DRILL 4:

1. Convert 4/5 to a percent.
2. Convert 3/6 to a percent.
3. Convert 9/12 to a percent.
4. Convert 6/20 to a percent.
5. Convert 3/2 to a percent.

DRILL SET 6:

DRILL 1

1. Forty percent of the balls in a bag are red. If there are a total of 300 balls in the bag, how many of them are red?

2. Anna always puts 15% of her salary in a retirement fund. If she put $6,000 in her retirement fund last year, what was her salary that year?

3. Sixteen students wear glasses. If there are forty students total, what fraction of the students wear glasses?

4. What is the new price of an eighty-dollar sweater after being discounted 20%?

5. Billy has twenty dollars and Johnny has thirty dollars. If both Billy and Johnny invest 25% of their combined money in baseball cards, how much money will they invest?

DRILL 2

1. Last year, a furniture store sold four hundred chairs, two hundred tables, four hundred couches, and nothing else. The chairs made up what percent of the total items sold?

2. At her old job, Marie earned a yearly salary of $80,000. At her new job, Marie earns a salary equal to $40,000 plus a commission of 25% on all her sales. If Marie wants to make a yearly salary at her new job that is the same as that of her old job, how much does she have to produce in terms of yearly sales?

3. Of 250 people surveyed, 36% said they preferred regular soda, 40% said they preferred diet soda, and the rest did not have a preference. How many of the 250 people did not have a preference?

4. A jar contains 1/3 red marbles and 1/2 blue marbles. The remaining 25 marbles are white. How many marbles does the jar contain?

5. Ted has 2/3 as many friends as Billy has, and Chris has 1/2 as many friends as Billy has. The number of friends that Chris has is what percent of the number of friends that Ted has?

Drill Set Answers

Set 1, Drill 1

1. $\mathbf{\frac{3}{4}}$: The denominators are the same, but the numerator of $\frac{3}{4}$ is larger, so $\frac{3}{4} > \frac{1}{4}$.

2. $\mathbf{\frac{1}{5}}$: The numerators are the same, but the denominator of $\frac{1}{5}$ is smaller, so $\frac{1}{5} > \frac{1}{6}$.

3. $\mathbf{\frac{53}{52}}$: In the first fraction, $\frac{53}{52}$, the numerator is bigger than the denominator, so the fraction is greater than 1. In the second fraction, $\frac{85}{86}$, the denominator is bigger than the numerator, so the fraction is less than 1. Thus, $\frac{53}{52} > \frac{85}{86}$.

4. $\mathbf{\frac{7}{9}}$: The second fraction, $\frac{6}{10}$, has both a smaller numerator and a larger denominator than the first fraction. Therefore, $\frac{6}{10} < \frac{7}{9}$.

5. $\mathbf{\frac{590}{290}}$: The first fraction is greater than 1 but less than 2, because 700 is less than twice 360 ($2 \times 360 = 720$). The second fraction is greater than 2, because 590 is more than twice 290 ($2 \times 290 = 580$). Thus, $\frac{590}{290} > \frac{700}{360}$.

Set 1, Drill 2

1. $\mathbf{\frac{5}{7}}$: $\frac{2}{7} + \frac{3}{7} = \frac{2+3}{7} = \frac{5}{7}$

2. $\mathbf{\frac{1}{8}}$: $\frac{5}{8} - \frac{4}{8} = \frac{5-4}{8} = \frac{1}{8}$

3. $\mathbf{\frac{5}{9}}$: $\frac{7}{9} - \frac{2}{9} = \frac{7-2}{9} = \frac{5}{9}$

4. $\mathbf{\frac{29}{11}}$: $\frac{9}{11} + \frac{20}{11} = \frac{9+20}{11} = \frac{29}{11}$

5. $\mathbf{\frac{-7}{4}}$: $\frac{3}{4} - \frac{10}{4} = \frac{3-10}{4} = \frac{-7}{4}$

Set 1, Drill 3

1. $\mathbf{\frac{11}{9}}$: $\frac{2}{3} + \frac{5}{9} = \frac{2}{3} \times \frac{3}{3} + \frac{5}{9} = \frac{2\times 3}{3\times 3} + \frac{5}{9} = \frac{6}{9} + \frac{5}{9} = \frac{6+5}{9} = \frac{11}{9}$

2. $\mathbf{\frac{-3}{8}}$: $\frac{7}{8} - \frac{5}{4} = \frac{7}{8} - \frac{5}{4} \times \frac{2}{2} = \frac{7}{8} - \frac{5\times 2}{4\times 2} = \frac{7}{8} - \frac{10}{8} = \frac{7-10}{8} = \frac{-3}{8}$

3. $\mathbf{\frac{116}{99}}$: $\frac{4}{9} + \frac{8}{11} = \frac{4}{9} \times \frac{11}{11} + \frac{8}{11} \times \frac{9}{9} = \frac{4\times 11}{9\times 11} + \frac{8\times 9}{11\times 9} = \frac{44}{99} + \frac{72}{99} = \frac{116}{99}$

4. **0**: $\frac{20}{12}-\frac{5}{3}=\frac{20}{12}-\frac{5}{3}\times\frac{4}{4}=\frac{20}{12}-\frac{5\times4}{3\times4}=\frac{20}{12}-\frac{20}{12}=0$

5. $\mathbf{\frac{67}{40}}$: $\frac{1}{4}+\frac{4}{5}+\frac{5}{8}=\frac{1}{4}\times\frac{10}{10}+\frac{4}{5}\times\frac{8}{8}+\frac{5}{8}\times\frac{5}{5}=\frac{1\times10}{4\times10}+\frac{4\times8}{5\times8}+\frac{5\times5}{8\times5}=$

$\frac{10}{40}+\frac{32}{40}+\frac{25}{40}=\frac{10+32+25}{40}=\frac{67}{40}$

Set 1, Drill 4

1. 3: $\frac{1}{5}+\frac{x}{5}=\frac{4}{5}$

$\frac{x}{5}=\frac{4}{5}-\frac{1}{5}$

$\frac{x}{5}=\frac{4-1}{5}=\frac{3}{5}$

$x=3$

2. 13: $\frac{x}{8}-\frac{3}{8}=\frac{10}{8}$

$\frac{x}{8}=\frac{10}{8}+\frac{3}{8}$

$\frac{x}{8}=\frac{10+3}{8}=\frac{13}{8}$

$x=13$

3. 3: $\frac{x}{6}+\frac{5}{12}=\frac{11}{12}$

$\frac{x}{6}=\frac{11}{12}-\frac{5}{12}$

$\frac{x}{6}=\frac{11-5}{12}=\frac{6}{12}$

$\frac{x\times2}{6\times2}=\frac{2x}{12}=\frac{6}{12}$

$2x=6$

$x=3$

4. 8: $\frac{2}{7}-\frac{x}{21}=-\frac{2}{21}$

$-\frac{x}{21}=-\frac{2}{21}-\frac{2}{7}$

$-\frac{x}{21}=-\frac{2}{21}-\frac{2\times3}{7\times3}$

$-\frac{x}{21}=-\frac{2}{21}-\frac{6}{21}=\frac{-2-6}{21}=\frac{-8}{21}$

$\frac{x}{21}=\frac{8}{21}$

$x=8$

5. **3:** $\frac{2}{5}+\frac{x}{8}=\frac{31}{40}$

$\frac{x}{8}=\frac{31}{40}-\frac{2}{5}$

$\frac{x}{8}=\frac{31}{40}-\frac{2\times8}{5\times8}=\frac{31}{40}-\frac{16}{40}=\frac{15}{40}$

$\frac{x\times5}{8\times5}=\frac{5x}{40}=\frac{15}{40}$

$5x=15$

$x=3$

Set 2, Drill 1

1. $\mathbf{\frac{2}{3}}: \frac{6}{9} = \frac{2\times3}{3\times3} = \frac{2}{3}\times\frac{3}{3} = \frac{2}{3}$

2. $\mathbf{\frac{3}{7}}: \frac{12}{28} = \frac{2\times2\times3}{2\times2\times7} = \frac{3}{7}\times\frac{2\times2}{2\times2} = \frac{2}{3}$

3. $\mathbf{\frac{2}{3}}: \frac{24}{36} = \frac{2\times2\times2\times3}{2\times2\times3\times3} = \frac{2}{3}\times\frac{2\times2\times3}{2\times2\times3} = \frac{2}{3}$

4. $\mathbf{\frac{7}{200}}: \frac{35}{100} = \frac{5\times7}{2\times2\times5\times5} = \frac{7}{2\times2\times5}\times\frac{5}{5} = \frac{7}{20}$

5. $\mathbf{\frac{3x}{35}}: \frac{6x}{70} = \frac{2\times3\times x}{2\times5\times7} = \frac{3\times x}{5\times7}\times\frac{2}{2} = \frac{3x}{35}$

Set 2, Drill 2

1. $\mathbf{\frac{32}{63}}: \frac{4}{7}\times\frac{8}{9} = \frac{4\times8}{7\times9} = \frac{32}{63}$

2. $\mathbf{\frac{27}{8}}: \frac{9}{4}\div\frac{2}{3} = \frac{9}{4}\times\frac{3}{2} = \frac{9\times3}{4\times2} = \frac{27}{8}$

3. $\mathbf{\frac{7}{24}}: \frac{7}{15}\div\frac{8}{5} = \frac{7}{15}\times\frac{5}{8} = \frac{7\times5}{3\times5\times8} = \frac{7\times\cancel{5}}{3\times\cancel{5}\times8} = \frac{7}{24}$

4. $\mathbf{\frac{1}{3}}: \frac{5}{12}\times\frac{8}{10} = \frac{5\times2\times2\times2}{2\times2\times3\times2\times5} = \frac{\cancel{5}\times\cancel{2}\times\cancel{2}\times\cancel{2}}{\cancel{2}\times\cancel{2}\times3\times\cancel{2}\times\cancel{5}} = \frac{1}{3}$

5. $\mathbf{\frac{5x}{26}}: \frac{3x}{13}\times\frac{5}{6} = \frac{3\times x\times5}{13\times2\times3} = \frac{\cancel{3}\times x\times5}{13\times2\times\cancel{3}} = \frac{5x}{26}$

Set 2, Drill 3

1. $\mathbf{\frac{1}{2}}: \frac{14}{20}\times\frac{15}{21} = \frac{2\times7\times3\times5}{2\times2\times5\times7\times3} = \frac{\cancel{2}\times\cancel{7}\times\cancel{3}\times\cancel{5}}{2\times\cancel{2}\times\cancel{5}\times\cancel{7}\times\cancel{3}} = \frac{1}{2}$

2. $\mathbf{\frac{4}{15}}: \frac{6}{25}\div\frac{9}{10} = \frac{6}{25}\times\frac{10}{9} = \frac{2\times3\times2\times5}{5\times5\times3\times3} = \frac{2\times\cancel{3}\times2\times\cancel{5}}{5\times\cancel{5}\times\cancel{3}\times3} = \frac{4}{15}$

3. $\mathbf{\frac{5}{39}}: \frac{4}{21}\times\frac{14}{13}\times\frac{5}{8} = \frac{2\times2\times2\times7\times5}{3\times7\times13\times2\times2\times2} = \frac{\cancel{2}\times\cancel{2}\times\cancel{2}\times\cancel{7}\times5}{3\times\cancel{7}\times13\times\cancel{2}\times\cancel{2}\times\cancel{2}} = \frac{5}{39}$

4. **1:** $\frac{3}{11}\div\frac{3}{11} = \frac{3}{11}\times\frac{11}{3} = \frac{3\times11}{11\times3} = \frac{33}{33} = 1$

5. **0:** $\frac{57}{63}\times\frac{0}{18} = \frac{57\times0}{63\times18} = 0$

Set 2, Drill 4

1. $\mathbf{\frac{9}{35}}$**:** $\frac{5}{3}x = \frac{3}{7}$

$$x = \frac{3}{7} \div \frac{5}{3} = \frac{3}{7} \times \frac{3}{5}$$

$$x = \frac{9}{35}$$

2. $\mathbf{\frac{6}{7}}$**:** $\frac{2}{x} = \frac{7}{3}$

$$2 \times 3 = 7 \times x$$

$$6 = 7x$$

$$\frac{6}{7} = x$$

3. $\mathbf{\frac{5}{9}}$**:** $\frac{6}{11}x = \frac{10}{33}$

$$x = \frac{10}{33} \div \frac{6}{11} = \frac{10}{33} \times \frac{11}{6}$$

$$x = \frac{2 \times 5 \times 11}{3 \times 11 \times 2 \times 3} = \frac{\cancel{2} \times 5 \times \cancel{11}}{3 \times \cancel{11} \times \cancel{2} \times 3}$$

$$x = \frac{5}{9}$$

4. $\mathbf{\frac{39}{14}}$**:** $\frac{2x}{13} = \frac{3}{7}$

$$2x \times 7 = 3 \times 13$$

$$14x = 39$$

$$x = \frac{39}{14}$$

5. **3:** $\frac{3x}{4} - \frac{5}{6} = \frac{17}{12}$

$$\frac{3x}{4} \times \frac{3}{3} - \frac{5}{6} \times \frac{2}{2} = \frac{17}{12}$$

$$\frac{9x}{12} - \frac{10}{12} = \frac{17}{12}$$

$$9x = 10 = 17$$

$$9x = 27$$

$$x = 3$$

Set 3, Drill 1

1. $\mathbf{2\frac{1}{4}}$: $\frac{9}{4}=\frac{8+1}{4}=\frac{8}{4}+\frac{1}{4}=2+\frac{1}{4}=2\frac{1}{4}$

2. $\mathbf{4\frac{3}{7}}$: $\frac{31}{7}=\frac{28+3}{7}=\frac{28}{7}+\frac{3}{7}=4+\frac{3}{7}=4\frac{3}{7}$

3. $\mathbf{3\frac{2}{15}}$: $\frac{47}{15}=\frac{45+2}{15}=\frac{45}{15}+\frac{2}{15}=3+\frac{2}{15}=3\frac{2}{15}$

4. $\mathbf{3\frac{1}{2}}$: $\frac{70}{20}=\frac{60+10}{20}=\frac{60}{20}+\frac{10}{20}=3+\frac{10}{20}=3+\frac{1}{2}=3\frac{1}{2}$

5. **7**: $\frac{91}{13}=7$

Set 3, Drill 2

1. $\mathbf{\frac{11}{3}}$: $3\frac{2}{3}=3+\frac{2}{3}=\frac{3\times3}{1\times3}+\frac{2}{3}=\frac{9}{3}+\frac{2}{3}=\frac{11}{3}$

2. $\mathbf{\frac{13}{6}}$: $2\frac{1}{6}=2+\frac{1}{6}=\frac{2\times6}{1\times6}+\frac{1}{6}=\frac{12}{6}+\frac{1}{6}=\frac{13}{6}$

3. $\mathbf{\frac{45}{7}}$: $6\frac{3}{7}=6+\frac{3}{7}=\frac{6\times7}{1\times7}+\frac{3}{7}=\frac{42}{7}+\frac{3}{7}=\frac{45}{7}$

4. $\mathbf{\frac{41}{9}}$: $4\frac{5}{9}=4+\frac{5}{9}=\frac{4\times9}{1\times9}+\frac{5}{9}+\frac{36}{9}+\frac{5}{9}=\frac{41}{9}$

5. $\mathbf{\frac{149}{12}}$: $12\frac{5}{12}=12+\frac{5}{12}=\frac{12\times12}{1\times12}+\frac{5}{12}=\frac{144}{12}+\frac{5}{12}=\frac{149}{42}$

Set 4, Drill 1

1. $6.75\times10^{3}=$ **6,750** — Move the decimal to the right 3 places.
2. $72.12\times10^{-4}=$ **0.007212** — Move the decimal to the left 4 places
3. $2{,}346\times10^{-3}=$ **2.346** — Move the decimal to the left 3 places.
4. $27.048\times10^{2}=$ **2,704.8** — Move the decimal to the right 2 places.
5. $54.197\ /\ 10^{2}=$ **0.54197** — Because we are dividing by 10^2, we move the decimal to the left 2 places.

Set 4, Drill 2

1.
$$\begin{array}{r} 1.000 \\ +\,0.200 \\ +\,0.030 \\ +\,0.004 \\ \hline 1.234 \end{array}$$

2.
$$\begin{array}{r} 0.\overset{1}{4}8 \\ +\,0.02 \\ \hline 0.50 \end{array}$$

3.
$$\begin{array}{r} 1.21 \\ +\,0.38 \\ \hline 1.59 \end{array}$$

4. $$\begin{array}{r} 0.35 \\ -\,0.02 \\ \hline 0.33 \end{array}$$

5. $$\begin{array}{r} 0.\overset{1}{3}70 \\ +\,0.042 \\ \hline 0.412 \end{array}$$

Set 4, Drill 3

1. **0.54:** $0.27 \times 2 =$
   $27 \times 2 = 54$ Move the decimal to the left 2 places.
   $0.27 \times 2 = 0.54$

2. **1.209:** $0.403 \times 3 =$
   $403 \times 3 = 1{,}209$ Move the decimal to the left 3 places.
   $0.403 \times 3 = 1.209$

3. **0.04:** $0.2 \times 0.2 =$
   $2 \times 2 = 4$ Move the decimal to the left 2 places.
   $0.2 \times 0.2 = 0.04$

4. **17:** $20 \times 0.85 =$
   $20 \times 85 = 1{,}700$ Move the decimal to the left 2 places.
   $20 \times 0.85 = 17$

5. **0.00804:** $0.04 \times 0.201 =$
   $4 \times 201 = 804$ Move the decimal to the left 5 places.
   $0.04 \times 0.201 = 0.00804$

Set 4, Drill 4

1. **0.168:** $2.1 \times 0.08 =$
   $21 \times 8 = 168$ Move the decimal to the left 3 places.
   $2.1 \times 0.08 = 0.168$

2. **0.0252:** $0.063 \times 0.4 =$
   $63 \times 4 = 252$ Move the decimal to the left 4 places.
   $0.063 \times 0.4 = 0.0252$

3. **0.00015:** $0.03 \times 0.005 =$
   $3 \times 5 = 15$ Move the decimal to the left 5 places.
   $0.03 \times 0.005 = 0.00015$

4. **70:** $0.75(80) + 0.50(20) =$ Break this problem into two multiplication problems
   $0.75 \times 80 = 60$
   $0.50 \times 20 = 10$ Now add the two results.
   $60 + 10 = 70$

5. **0.01:** $100 \times 0.01 \times 0.01 =$
   $100 \times 1 \times 1 = 100$ Move the decimal to the left 4 places.
   $100 \times 0.01 \times 0.01 = 0.01$

Set 4, Drill 5

1. **20** : $\frac{4}{0.2} =$

$$\frac{4}{0.2} \times \frac{10}{10} = \frac{40}{2} = 20$$

2. **42** : $\frac{12.6}{0.3} =$

$$\frac{12.6}{0.3} \times \frac{10}{10} = \frac{126}{3} = 42$$

3. **80** : $\frac{3.20}{0.04} =$

$$\frac{3.20}{0.04} \times \frac{100}{100} = \frac{320}{4} = 80$$

4. **7** : $\frac{0.49}{0.07} =$

$$\frac{0.49}{0.07} \times \frac{100}{100} = \frac{49}{7} = 7$$

5. **12** : $\frac{6}{0.5} =$

$$\frac{6}{0.5} \times \frac{10}{10} = \frac{60}{5} = 12$$

Set 5, Drill 1

Fraction	Decimal	Percent
1/100	0.01	1%
1/20	0.05	5%
1/10	0.1	10%
1/8	0.125	12.5%
1/5	0.2	20%
1/4	0.25	25%
3/10	0.3	30%
1/3	0.3333...	33.33...%
3/8	0.375	37.5%
2/5	0.40	40%
1/2	0.50	50%
3/5	0.6	60%
2/3	0.6666...	66.66...%
7/10	0.7	70%
3/4	0.75	75%

Fraction	Decimal	Percent
4/5	0.8	80%
7/8	0.875	87.5%
9/10	0.9	90%
1	1.0	100%

Set 5, Drill 2

1. **0.45:** 45% becomes 0.45.
2. **7/10:** 70% becomes 70/100, which reduces to 7/10.
3. **0.1325:** 13.25% becomes 0.1325.
4. **9/25:** 36% becomes 36/100, which reduces to 9/25.
5. **0.0002:** 0.02% becomes 0.0002.

Set 5, Drill 3

1. **20%:** 0.20 becomes 20%.
2. **11/20:** 0.55 becomes 55/100, which reduces to 11/20.
3. **30.4%:** 0.304 becomes 30.4%.
4. **91/200:** 0.455 becomes 455/1000, which reduces to 91/200.
5. **3/8:** 0.375 becomes 375/1000, which reduces to 3/8.

Set 5, Drill 4

1. **80%:** Step 1: $4 \div 5 = 0.8$
   Step 2: 0.8 becomes 80%
   $5\overline{)4.0}$ = 0.8

2. **50%:** Step 1: $3 \div 6 = 0.5$
   Step 2: 0.5 becomes 50%
   $6\overline{)3.0}$ = 0.5

3. **75%:** Step 1: $9 \div 12 = 0.75$
   Step 2: 0.75 becomes 75%
   $12\overline{)9.00}$ = 0.75

4. **30%:** Step 1: $6 \div 20 = 0.30$
   Step 2: 0.30 becomes 30%
   $20\overline{)6.0}$ = 0.3

5. **150%:** Step 1: $3 \div 2 = 1.5$
   Step 2: 1.5 becomes 150%
   $2\overline{)3.0}$ = 1.5

Set 6, Drill 1

1. **120:** $300 \times 40\% = 300 \times 4/10 = 120$.

2. **$40,000:** Let a = Anna's total salary
   $\$6{,}000 = 15\%$ of a
   $\$6{,}000 = 15/100 \times a$
   $\$6{,}000 = 3/20 \times a$
   $\$6{,}000 \times 20/3 = a$
   $\$40{,}000 = a$

3. **2/5:** $\frac{16}{40} = \frac{2\times2\times2\times2}{2\times2\times2\times5} = \frac{2}{5}$.

4. **$64:** If the sweater has been discounted 20%, then the new price is 80% of the original (because 100% – 20% = 80%):

   80% × $80 = 4/5 × $80 = $64

5. **$12.50:** (0.25 × $20) + (0.25 × $30) = $5 + $7.50 = $12.50

Set 6, Drill 2

1. **40%:** 400/(400 + 200 + 400) = 400/1000 = 4/10 = 40%

2. **$160,000:** Let m represent the amount of sales that Marie needs to generate in her new job to equal her previous salary:

   $80,000 = $40,000 + m × 25%
   $80,000 = $40,000 + 1/4 m
   $40,000 = 1/4 m
   $160,000 = m

3. **60:** The percent of people who did not have a preference = 100% – (36% + 40%) = 100% – 76% = 24%

   The number of people who did not have a preference = 24% × 250 =
   $\frac{24}{100}\times 250 = \frac{6}{25}\times 250 = \frac{6}{\cancel{25}}\times\frac{\cancel{25}\times 10}{1} = 60$

4. **150:** First we need to figure out what fraction of the marbles are white. We can do this by figuring out what fraction of the marbles are not white. Add the fractional amounts of red and blue marbles.

   1/3 + 1/2 = 2/6 + 3/6 = 5/6.

   1 – 5/6 = 1/6 The white marbles are 1/6 of the total number of marbles.

   Let x = total number of marbles:
   25 = 1/6 x
   150 = x

5. **75%:** Let T = the number of friends Ted has.
   Let B = the number of friends Billy has.
   Let C = the number of friends Chris has.

   $T = 2/3\ B$
   $C = 1/2\ B$

   $C = x\%$ of T?

   $B = 3/2\ T$
   $C = 1/2\ (3/2\ T)$
   $C = 3/4\ T$
   $C = 75\%\ T$

   Thus Chris has 75 percent as many friends as Tim has.

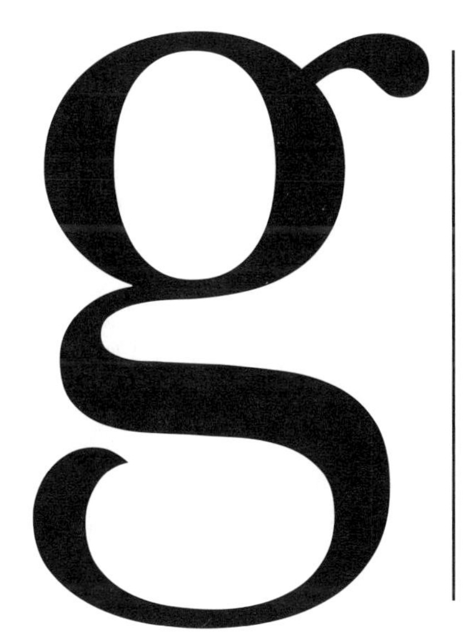

Chapter 7
of
FRACTIONS, DECIMALS, & PERCENTS

FDPs PRACTICE QUESTION SETS

In This Chapter . . .

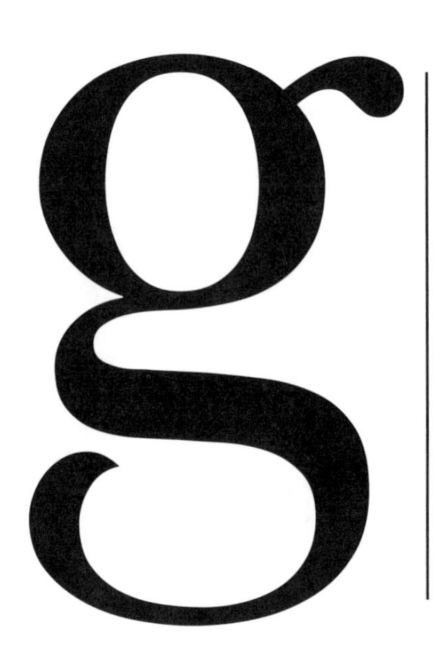

- Easy Practice Question Set
- Medium Practice Question Set
- Hard Practice Question Set
- Easy Practice Question Solutions
- Medium Practice Question Solutions
- Hard Practice Question Solutions

FDPs: Easy Practice Question Set

1. If $\frac{1}{2}$ of the students in a class are women and $\frac{3}{4}$ of these women are employed full time, what portion of the class consists of women that are not employed full time?

   (A) $\frac{1}{8}$
   (B) $\frac{3}{8}$
   (C) $\frac{1}{2}$
   (D) $\frac{5}{8}$
   (E) $\frac{7}{8}$

2.

Quantity A	Quantity B
$\frac{1}{3} \div \frac{4}{5}$	$\frac{1}{3} \times \frac{4}{5}$

3.

Quantity A	Quantity B
The tens digit of (6.4×3)	The thousandths digit of $\frac{3}{8}$

4. If $35x$ is 50% of 7, $x =$

   (A) 0.1
   (B) 0.5
   (C) 1
   (D) 5
   (E) 7

5. If 25 is less than 10% of x, x could be which of the following?

   Indicate all such values.

   [A] 2.5
   [B] 25
   [C] 125
   [D] 250
   [E] 300
   [F] 500

6. $\frac{1}{3}$ of the 90 ducks on a pond are mallards. If 22 ruddy ducks and 18 white–faced whistling ducks depart from the pond, what fraction of the remaining ducks are mallards?

(A) $\frac{3}{13}$
(B) $\frac{1}{3}$
(C) $\frac{4}{9}$
(D) $\frac{3}{5}$
(E) $\frac{2}{3}$

7. If John has $100 to spend and was unable to buy a particular item before the price was reduced by 25%, but can afford the item after the price reduction, what was the maximum possible price of the item prior to the price reduction?

(A) $110.50
(B) $120.66
(C) $125.00
(D) $133.33
(E) $140.00

8. What is the value of $(3.6 \times 0.2 - 0.3) \div 0.02$?

(A) –18
(B) –1.8
(C) 1.8
(D) 2.1
(E) 21

9.

a is the hundredths digit of *b*, which is equal to 1.5*a*4.

Quantity A	Quantity B
b rounded to the nearest hundredth	*b* rounded to the nearest tenth

10.

Quantity A	Quantity B
$\frac{3}{4} - \frac{1}{3}$	$\frac{3}{4} \times \frac{1}{3}$

11.

p is a fraction between 0 and 1.

Quantity A	**Quantity B**
The average of p and its reciprocal	1

12.

After a 20% discount, a dress is sold at a price of $176.

Quantity A	**Quantity B**
The original price of the dress	$200

13. The average home price in Pleasantville was $260,000 in 2008. If the average home price rose by 5% from 2008 to 2009, and then fell 10% from 2009 to 2010, what was the average home price in 2010?

(A) $222,300
(B) $245,700
(C) $247,000
(D) $271,000
(E) $300,300

14.

Quantity A	**Quantity B**
The value of $1,000 after earning simple interest at 20% for 5 years	The value of $1,000 after earning compound interest at 20% for 4 years, compounded annually

15. 150% of $\frac{3}{2}$ equals:

(A) $\frac{2}{3}$

(B) $\frac{3}{4}$

(C) 1

(D) $\frac{9}{4}$

(E) $\frac{15}{4}$

16.

Quantity A	Quantity B
$\frac{3}{8}$ of 288	110

17. According to a survey, $\frac{3}{5}$ of students at University X eat lunch at the university's dining hall, while only $\frac{1}{6}$ of the university's faculty members eat lunch at the dining hall.

Quantity A	Quantity B
The number of students and faculty members who eat lunch at the university's dining hall	The number of students and faculty members who do *not* eat lunch at the university's dining hall

18.

15 is $\frac{5}{3}$ of x.

Quantity A	Quantity B
x	10

19. A comet is traveling towards a distant planet at a speed of 85,000 kilometers per minute. If the comet is currently 17 billion kilometers from the planet, approximately how many hours will it take for the comet to hit the planet?

(A) 350
(B) 1,333
(C) 2,000
(D) 2,500
(E) 3,333

20.

Quantity A	Quantity B
$\frac{749}{299}$	250%

FDPs: Medium Practice Question Set

1. A parking lot has a number of automobiles, each of which is a car or a truck. If there are 3 times as many cars as trucks and 20% of the cars are foreign, what percentage of the automobiles are foreign cars?

    (A) 7.5%
    (B) 12%
    (C) 13 1/3%
    (D) 15%
    (E) 20%

2. Suppose that x and y are integers and that $0 < x < y < 10$. The tenths digit of the decimal representation of $\frac{x}{16}$ is 5. What is the hundredths digit of the result of $\frac{17}{y}$?

    [ ]

3. How many one-fourths are in $\frac{3}{5}$ of $12\frac{1}{2}$?

    [ ]

4. What percent of 6 is equal to 15 percent of 150 percent of 20?

    Give your answer in percentage points: [ ] %

5. 20% of the professors at a symposium are psychologists, 60% are biologists, and the remaining 12 professors are economists. If 20 of the professors wear glasses, what percent of the professors do not wear glasses?

    (E) 20%
    (F) $33\frac{1}{3}\%$
    (G) 50%
    (H) $66\frac{2}{3}\%$
    (E) 80%

6.

Quantity A	Quantity B
The hundreds digit of $5^3 10^2$	The hundredths digit of $\frac{1}{4}$

7.

Year	Percent change from previous year (rounded to nearest 0.1 percent)
2006	–8.8%
2007	7.5%
2008	2.0%
2009	–4.9%

The annual percent change in the number of babies born at Hospital X is given in the table above. If 102 babies were born at Hospital X in 2005, how many babies were born in Hospital X in 2009?

[ ] babies

8.

x is a positive integer.

Quantity A	Quantity B
$\dfrac{\frac{5x}{7}}{\frac{8}{2x+3}}$	$\dfrac{x^2}{4}+\dfrac{8x}{7}$

9. What is the result of the following calculation, when expressed as a simple fraction?

$$\frac{3\frac{1}{4}-2\frac{2}{3}}{1\frac{1}{2}}=$$

Express your answer as a fraction: $\dfrac{[\ ]}{[\ ]}$

10.

Quantity A	Quantity B
$\dfrac{\frac{3}{2}+\frac{5}{6}}{\frac{4}{2}+\frac{4}{6}}$	$\dfrac{7}{8}$

11. The ratio of $\frac{2}{5}$ of 4 to $\frac{4}{3}$ of 5 is equal to the ratio of:

(I) 16 to 3
(J) 2 to 3
(K) 6 to 25
(L) 8 to 15
(E) 6 to 1

12. $\frac{3}{5}$ of the workers in a department are men, and the rest are women. If $\frac{1}{2}$ of the men and $\frac{3}{7}$ of the women in the department are over 35, what fraction of all the workers in the department are over 35?

Express your answer as a fraction: $\frac{\square}{\square}$

13.

$q > 0$

Quantity A	Quantity B
The result of increasing, then decreasing, q by 1%	The result of increasing, then decreasing, q by 2%

14.

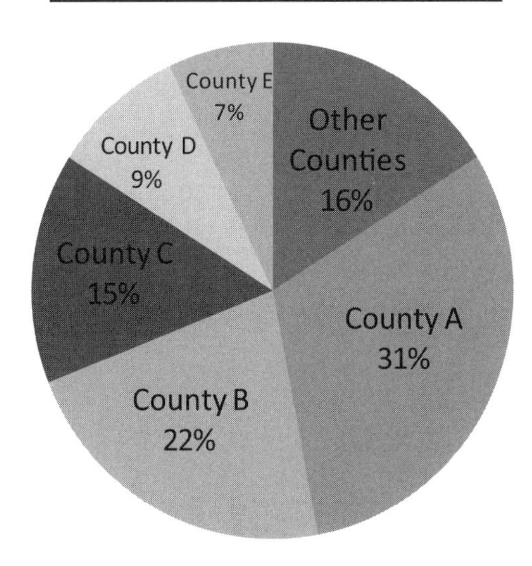

To the nearest percent, the combined population of the two most populous counties in State X is what percent greater than the combined population of all other counties?

(A) 3
(B) 6
(C) 13
(D) 35
(E) 113

15. If $x = 80\%$, by what percent is x larger than x^2?

    (A) 8%
    (B) 16%
    (C) 20%
    (D) 25%
    (E) 80%

16. Which of the following is equal to $\dfrac{\frac{3}{8} \div \left(\frac{3}{4} - \frac{1}{3}\right)}{\frac{2}{5}}$?

    (A) 2.25
    (B) 1.25
    (C) $1.60\overline{714285}$
    (D) 0.390625
    (E) 0.078125

17. A bar over a sequence of digits in a decimal indicates that the sequence repeats indefinitely. Which of the following fractions is equivalent to $0.\overline{12}$?

    (A) $\frac{3}{25}$
    (B) $\frac{4}{33}$
    (C) $\frac{109}{900}$
    (D) $\frac{11}{90}$
    (E) $\frac{12}{90}$

18. $\dfrac{(1)(0.1)+(2)(0.2)+(3)(0.3)+(4)(0.4)}{(1)(0.1)+\left(\frac{1}{2}\right)(0.2)+\left(\frac{1}{3}\right)(0.3)+\left(\frac{1}{4}\right)(0.4)} =$

    (A) 2.5
    (B) 3
    (C) 7.5
    (D) 10
    (E) 30

19. $\dfrac{\dfrac{5}{3}-\dfrac{1}{6}}{\dfrac{0.3^2}{0.2}}=$

(A) $\dfrac{1}{30}$

(B) $\dfrac{1}{3}$

(C) $\dfrac{10}{3}$

(D) $\dfrac{25}{3}$

(E) $\dfrac{5}{6}$

20. A shipping company charges a shipping fee that varies depending upon the weight of the package to be shipped. The price to ship all packages weighing below or equal to 5 pounds is $2.80, and an additional $0.25 is charged for each additional pound above 5 pounds. If the cost to ship Package A is greater than $7 but less than $8, which of the following could be the weight of Package A, in pounds?

Indicate all that apply.

[A] 20

[B] 21

[C] 22

[D] 23

[E] 24

[F] 25

[G] 26

FDPs: Hard Practice Question Set

CAUTION: These problems are *very difficult*—more difficult than many of the problems you will likely see on the GRE. Consider these "Challenge Problems." Have fun!

1.

$x > 4$

Quantity A	Quantity B
$\left[\dfrac{\frac{(x^2-9)}{3}}{\frac{x+3}{8}}\right]^{-1}$	$\dfrac{3}{8}$

2. Suppose $x_1 = 2 \times 10^5$, $x_2 = 2^5 \times 5^5$, $y_1 = 200$, and $y_2 = 400$. x_1 and x_2 and represent the value of x in years 1 and 2, respectively, and y_1 and y_2 and represent the value of y in years 1 and 2, respectively.

Quantity A	Quantity B
The percent <u>decrease</u> in x from year 1 to year 2	The percent <u>increase</u> in y from year 1 to year 2

3.

x is a positive integer.

Quantity A	Quantity B
The units digit of 6^x	The units digit of 4^{2x}

4. If $x > 0$ and the quantities $\sqrt{x}$ and $\dfrac{\sqrt{x}}{x}$ are reciprocals, which of the following <u>could</u> be true?

Indicate <u>all</u> choices that apply.

(A) $\sqrt{x}$ is greater than $\dfrac{\sqrt{x}}{x}$

(B) $\sqrt{x}$ is less than $\dfrac{\sqrt{x}}{x}$

(C) $\sqrt{x}$ is greater than x

5. x is at least 50% greater than 100, and at most 100% greater than 100. y is between 20% and 50% of x, inclusive. Which of the following are possible values for y?

   Indicate all such possible values.

   [A] 10
   [B] 25
   [C] 40
   [D] 60
   [E] 90
   [F] 100
   [G] 120
   [H] 170
   [I] 200

6. A shopkeeper purchases birdfeeders for $10 each and sells them for $18 each. If the cost of the feeders increases by 50% for two months in a row, what is the smallest percent increase the shopkeeper can apply to the selling price in order to avoid selling at a loss?

   Give your answer in percentage points: [ ] %

7.

The integers x and $(x - 1)$ are not divisible by 4.

Quantity A	Quantity B
The value of the tenths digit of $\frac{x}{4}$	The value of the hundredths digit of $\frac{x}{4}$

8. What is the ones digit of $3^{23} - 2^{18}$?

   [ ]

9. In 1990, the annual salaries of clerks in Town Q ranged from $40,000 to $55,000. In 2000, the same group of salaries ranged from $70,000 to $110,000. The percent increase in the annual salary of a clerk in Town Q between 1990 and 2000 COULD be which of the following?

   Select all that apply.

   [A] 15
   [B] 50
   [C] 100
   [D] 150
   [E] 200

10. Last year, $\frac{3}{8}$ of the 420 Juniors at Central High School took French, $\frac{2}{5}$ took Geography, and $\frac{1}{4}$ took both.

    Which of the following statements MUST be true?

    Select all statements that apply.

    [A] More students took French than Geography.
    [B] More students took both French and Geography than took neither.
    [C] The number of students that took Geography but not French was greater than 50.

11. Jennifer has 40% more stamps than Peter. However, if she gives 45 of her stamps to Peter, then Peter will have 10% more stamps than Jennifer. How many stamps did Jennifer begin with?

    (A) 140
    (B) 175
    (C) 200
    (D) 220
    (E) 245

12. A dealer buys a boat at auction and pays 10% below list price. He then sells the boat at a profit of 30% of his cost of the boat. At the same selling price, what would the dealer's percent profit over his cost have been if he had bought the boat at list price?

    Round your answer to the nearest whole percentage point before entering: [ ] %

13. If $P > 0$ and $P\%$ of $3P$ is $P\%$ less than P, then P equals:

    (A) 5
    (B) 25
    (C) 40
    (D) 50
    (E) 64

14. To the nearest dollar, how much money needs to be invested in an account that earns 8% interest, compounded quarterly, in order to have $12,000 at the end of 3 years?

    (A) $9,462
    (B) $9,526
    (C) $9,677
    (D) $9,804
    (E) $11,308

15. At the beginning of each year Jane puts $2,000 in an account that earns 6% interest, compounded annually. To the nearest dollar, how much money will Jane have in the account at the end of 4 years if she makes no withdrawals?

    Round your answer to the nearest whole number: $ [ ]

16. How many zeros will the decimal equivalent of $\frac{1}{2^{11}5^{7}}+\frac{1}{2^{7}5^{11}}$ have after the decimal point prior to the first non-zero digit?

(A) 6
(B) 7
(C) 8
(D) 11
(E) 18

17. Which of the following is 10% less than 2?

(A) $\frac{1}{2}+1\frac{2}{5}$
(B) $\frac{5}{6}+1\frac{1}{3}$
(C) $1\frac{2}{3}+\frac{3}{10}$
(D) $2\frac{1}{6}-\frac{2}{5}$
(E) $2\frac{1}{2}-\frac{7}{10}$

18. The length of a rectangle is 3 inches, plus or minus an amount no greater than $\frac{1}{2}$ inch, while its width is 2 inches, plus or minus an amount no greater than $\frac{1}{4}$ inch. Which of the following could be the percentage by which the actual area of the rectangle is greater than or less than its nominal area of 6 square inches?

Select all that apply.

[A] −30%
[B] − 25%
[C] + 25%
[D] + 30%

19. *A* is the tens digit and *B* is the ones digit of the product 12,345 × 6,789. What is *AB*?

(A) 0
(B) 5
(C) 6
(D) 10
(E) 25

20. Which of the following could be the units digit 98^x, where x is an integer greater than 1?

Indicate all such digits:

[A] 0
[B] 1
[C] 2
[D] 3
[E] 4
[F] 5
[G] 6
[H] 7
[I] 8
[J] 9

FDPs: Easy Practice Question Solutions

1. **A:** If $\frac{3}{4}$ of the women are employed full time, then $\frac{1}{4}$ of the women are not employed full time. Thus, if $\frac{1}{2}$ of the class are women, then $\frac{1}{4}$ of $\frac{1}{2}$ of the class is female and not employed full time. This yields $\frac{1}{4} \times \frac{1}{2} = \frac{1}{8}$.

Another way to solve this problem is to choose a Smart Number for the population of the class. Since the two fractions in the question have denominators of 2 and 4, a Smart Number to choose for this problem is 8. Let 8 be the total number of people in the class. Then, since half of the class consists of women, there are 4 women. And, since $\frac{3}{4}$ of these women work, there are 3 women employed full time. This also tells us that there is 1 woman who is not employed full time. Thus the fraction of the class that consists of women who are not employed full time is $\frac{1}{8}$.

2. **A**: To simplify Quantity B, multiply straight across:

$$\frac{1}{3} \times \frac{4}{5} = \frac{4}{15}$$

To simplify Quantity A, flip and multiply:

$$\frac{1}{3} \div \frac{4}{5} = \frac{1}{3} \times \frac{5}{4} = \frac{5}{12}$$

Since Quantity A has both a larger numerator and a smaller denominator than Quantity B, Quantity A is greater.

To be absolutely certain, we can find a common denominator and multiply both results by that denominator to more easily compare:

Quantity A: $\frac{5}{12}\left(\frac{5}{5}\right) = \frac{25}{60}$

Quantity B: $\frac{4}{15}\left(\frac{4}{4}\right) = \frac{16}{60}$

Again, **Quantity A is greater**.

3. **B**: $6.4 \times 3 = 19.2$, so the tens digit of Quantity A equals 1.

$\frac{3}{8} = 0.375$, so the thousandths digit of Quantity B is 5.

Therefore **Quantity B is greater**.

4. **A**: The information in the question can be translated as follows:

$35x = 0.5 \times 7$
$35x = 3.5$
$x = 0.1$

Thus Choice A is the correct answer.

5. **E and F**: The information given in the question can be translated as follows:

$25 < 0.1x$

$\frac{25}{0.1} < x$

$250 < x$

Since x must be greater than 250, only the last 2 values are possibilities for x.

6. **D:** $\frac{1}{3}$ of 90 ducks = 30, so there are 30 mallards. 22 + 18 = 40, so 40 ducks in total depart from the pond.

This leaves 90 – 40 = 50 ducks remaining on the pond.

Mallards now represent $\frac{30}{50}$, or $\frac{3}{5}$, of the ducks on the pond.

7. **D:** In order to solve for the maximum price of the item we must solve the following inequality, where x is the original value of the item: $x - (25\% \text{ of } x) \leq \$100$, or $x - 0.25x \leq \$100$. This can be simplified to:

$x - 0.25x \leq \$100$

$0.75x \leq \$100$

$x \leq \$\frac{100}{75}$

$x \leq \$133\frac{1}{3}$

Thus x must be less than $\$133\frac{1}{3}$, and the largest price smaller than that figure (in dollar and cents) is $133.33.

A quick note on why Choice C is a trap answer and is not correct: A 25% reduction in the price of an item that costs $125 would yield a reduction of greater than $25 (if the original price of the item were $100, a 25% reduction would yield an exact reduction of $25, but the greater price leads to a greater dollar reduction). This implies that the new price would be less than $100. (If $125 were the original price, a 25% reduction would lead to a new price of $125 – $31.25 = $93.75.)

8. **E:** Using the acronym PEMDAS to remember the proper order of operations, we first perform the operations inside the parentheses. We multiply 3.6 and 0.2 and *only then* subtract 0.3 from the result. Finally, we divide by 0.02 to obtain:

$$(3.6\times0.2-0.3)\div0.02=\frac{(0.72-0.3)}{0.02}=\frac{0.42}{0.02}=21$$

Using the Calculator (which will automatically perform the multiplication before the subtraction), the keying order would be as follows:

3.6 × 0.2 = – 0.3 = ÷ 0.02 =

9. **D**: To approximate a decimal to the nearest hundredth, we must keep the first two digits after the decimal point, rounding up if the following digit is 5 or greater. In this case that digit is 4, so no rounding up is performed:

1.5*a* | *b* rounded to the nearest tenth

Likewise, to approximate a decimal to the nearest tenth, we must keep the first digit after the decimal point, rounding up if the following digit is 5 or greater. In this case that digit is equal to *a*, which may or may not be 5 or greater. Thus we do not know whether Quantity B should be 1.5 or 1.6; accordingly; we cannot tell whether it would be less than, equal to, or greater than 1.5*a*. **We do not have enough information** to answer the question.

10. **A**: To perform the calculation in Quantity A, set up the fractions to have a common denominator, which in this case is 3 × 4 = 12:

$$\frac{3}{4}-\frac{1}{3}=\frac{9}{12}-\frac{4}{12}=\frac{5}{12} \qquad \frac{3}{4}\times\frac{1}{3}$$

Meanwhile, for Quantity B, multiply the fractions by multiplying the numerators and denominators separately:

$$\frac{5}{12} \qquad \frac{3}{4}\times\frac{1}{3}=\frac{3\times1}{4\times3}=\frac{3}{12}$$

Because the fractions in both quantities have the same denominator, the quantities are easy to compare: the one with the greater numerator is greater. Therefore **Quantity A is greater** in this problem.

Note that, ordinarily, we would have cancelled the 3s in the numerator and denominator of Quantity B prior to the multiplication, so as to reduce the math needed to perform the multiplication. However, in this case we left them in place, because by doing so we were able to obtain the same denominator in Quantity B as in Quantity A, making the subsequent comparison easier.

11. **A:** One approach (probably the easiest) is to test a couple of fractions to observe what happens. For example, if $p=\frac{1}{2}$, then Quantity A is equal to the average of $\frac{1}{2}$ and 2, which equals $\frac{\frac{1}{2}+2}{2}=1.25$. Similarly, if $p=\frac{2}{3}$, then Quantity A equals $\frac{\left(\frac{2}{3}+\frac{3}{2}\right)}{2}=\frac{\frac{4}{6}+\frac{9}{6}}{2}=\frac{13}{12}$, which (again) is greater than 1. While it is true that testing numbers is rarely 100% conclusive, we might feel quite confident in this instance that a similar result would hold for all fractions between 0 and 1.

To prove the result algebraically is more challenging. Suppose that $p=\frac{a}{b}$, where *a* and *b* are positive and *b* > *a* (so that $0 < p < 1$). Then, the difference between 1 and p is given by $1-\frac{a}{b}=\frac{b-a}{b}$. The difference between the reciprocal of p and 1, meanwhile, is given by $\frac{b}{a}-1=\frac{b-a}{a}$. Note that both differences are fractions with the same numerator. In this case the fraction with the smaller denominator will be larger. In this case, the difference between the reciprocal of p and 1 is greater than the difference between 1 and p. Put differently, on a number line, the reciprocal of p is farther to the right of 1 than p is to the left of 1. Therefore the average of p and its reciprocal must be greater than 1.

12. **B:** One way to approach this problem is to solve directly for the original price of the dress. Call that value P. Then, we could write $P-\frac{20}{100}\times P=\$176$, from which we could isolate P. (Note that the 20% discount is taken on the original value and not on the final value.) While this approach is certainly correct, it is not the quickest way to arrive at the answer, because the arithmetic involved is somewhat cumbersome. A better way is to assume that the original price is \$200 and see how the resulting final price compares to \$176. If the original price is \$200, the discount is $\frac{20}{100}\times\$200=\$40$. The dress would then be sold for \$200 – \$40 = 160. This value is lower than \$176. Therefore \$200 must be too low compared to the actual original price, and **Quantity A must be greater.** (Incidentally, the original price of the dress is \$220.)

13. **B:** The result of a percent change to a quantity can be determined by way of a multiplier: Suppose that the initial value is P, which then rises by $5\%=\frac{5}{100}=0.05$. The resulting new price is equal to the original price plus the increase:
$P + 0.05P = 1.05P$. Similarly, the multiplier corresponding to a 10% decrease is equal to $P - 0.1P = 0.9P$.

The multiplier concept can save us time in computing the result of individual percent changes, but becomes even more useful when dealing with successive percent changes, as in this problem. The original \$260,000 becomes (0.9) × (1.05) × (\$260,000) = \$245,700.

Note that the answer is *not* the same as what would be obtained by simply reducing \$260,000 by 5%. This is because the initial 5% increase is based on the original \$260,000, but the following 10% decrease is based on the 2009 price, which is higher.

14. **B:** Simple interest earned at 20% annually on a principal of \$1,000 yields \$200 per year. Thus, after 5 years, the value of the principal plus interest will be \$1,000 + 5 × \$200 = \$2,000. Meanwhile, the final value of principal P plus interest at a rate of r, compounded annually for n years, is given by the formula $F = P(1 + r)^n$, where r is expressed as the decimal equivalent of the annual percentage interest rate. In this case, we have:

$F = \$1{,}000(1 + 0.2)^4$, which equals approximately \$2,073.6.  Therefore **Quantity B is larger.**

The value of \$1,000 after earning simple interest at 20% for 5 years = \$2,000	The value of \$1,000 after earning compound interest at 20% for 5 years, compounded annually = \$2,073.6

(It should be noted that, as a practical matter, the fourth power of a number is computed most easily by squaring the number twice. Squaring is accomplished on the GRE onscreen Calculator by the keystroke sequence "×=". Thus, the fastest way to obtain the value of F in the formula above is to key in "$\underbrace{\underbrace{1.2\times=}_{\text{squared}}\times=}_{4^{\text{th}}\text{ power}}\times1{,}000$".)

15. **D:** $150\% = 1.5 = \frac{3}{2}$. Therefore, 150% of $\frac{3}{2}$ equals $\frac{3}{2}\times\frac{3}{2}=\frac{9}{4}$.

16. **B:** $\frac{3}{8}$ of 288 is equal to $\frac{3}{8}\times288=3\times36=108$. **Therefore Quantity B is greater.**

17. **D:** The problem tells us that $\frac{3}{5}$ of the students at University X eat lunch at the dining hall, while $\frac{1}{6}$ of the faculty members eat there. However, we do not know how many students there are relative to the number of faculty members at this university. For example, if there are 9 students for every faculty member, then among the students and faculty members combined, $\frac{9\left(\frac{3}{5}\right)+1\left(\frac{1}{6}\right)}{9+1}=\frac{\frac{27}{5}+\frac{1}{6}}{10}=\frac{\frac{27\times6+5\times1}{30}}{10}=\frac{167}{300}$, or more than half, eat at the dining hall. By contrast, if there are only 2 students for every faculty member, then $\frac{2\left(\frac{3}{5}\right)+1\left(\frac{1}{6}\right)}{2+1}=\frac{\frac{6}{5}+\frac{1}{6}}{3}=\frac{\frac{6\times6+5\times1}{30}}{3}=\frac{41}{90}$ of the combined population eats at the dining hall—i.e., less than half. **We do not have enough information** to determine which quantity is greater.

18. **B:** If 15 is $\frac{5}{3}$ of x, then $15=\frac{5}{3}x$, so $x=\frac{3}{5}\times15=3\times3=9$. Therefore **Quantity B is greater.**

19. **E:** The comet is currently 17 billion, or 17,000,000,000, kilometers from the planet. Since it is traveling at a speed of 85,000 kilometers per minute, it will reach the planet in $\frac{17{,}000{,}000{,}000}{85{,}000}=\frac{17{,}000{,}000}{85}=\frac{1{,}000{,}000}{5}=200{,}000$ minutes. Since there are 60 minutes per hour, the comet will reach the planet in $\frac{200{,}000}{60}=3{,}333\frac{1}{3}\approx3{,}333$ hours.

20. **A:** Perhaps the easiest way to compare these quantities is to convert Quantity B to a decimal (250% = 2.5), and to note that Quantity is very similar to $\frac{750}{300}$, which is equal to 2.5. Let's call that fraction, $\frac{750}{300}$, the "baseline" fraction for Quantity A in this comparison.

Relative to the baseline fraction, the actual fraction in Quantity A has a numerator that is 1 lower than 750, and a denominator that is 1 lower than 300. Notice that subtracting 1 from 750 takes away a smaller portion of the original (baseline) number than does subtracting 1 from 300. In other words, as a proportion of the original number in the "baseline" fraction, the decrease in the denominator is larger than the decrease in the numerator. Since the denominator is falling faster than the numerator, relative to the baseline fraction, the fraction in Quantity A should be slightly larger than the baseline fraction. Therefore, **Quantity A is greater** than Quantity B.

Notice that you can prove that $\frac{749}{299}$ is slightly greater than 2.5 by plugging it into the GRE onscreen Calculator. If you do this, you will see that $\frac{749}{299}\approx2.50517$, which again is just larger than Quantity B.

FDPs: Medium Practice Question Solutions

1. **D:** If there are 3 times as many cars as trucks, then for every truck that appears in the lot there are exactly 3 cars. That means that for every set of 4 automobiles, there will be exactly 3 cars, so cars make up $\frac{3}{4} = 75\%$ of the total population of the lot. If 20% of these cars are foreign, then we want 20% of 75%, which is (0.20)(0.75) = 0.15 = 15%.

We may also solve this problem by choosing numbers, and the best number to choose for the number of automobiles in the lot is 100, since we are dealing with percentages. If there are 100 automobiles in the lot and there are 3 times as many cars as trucks then there are 75 cars and 25 trucks. If 20% of the cars are foreign, then (20%)(75) = (0.20)(75) = 15 cars are foreign. Therefore, the percentage of foreign cars in the lot of automobiles is given by $\frac{15}{100} =$ 15%.

2. **8:** Since the tenths digit of the decimal version of $\frac{x}{16}$ is 5, x must have a value of at least half of 16. That is, $x \geq 8$. Since $y > x$ and $y < 10$, and both must be integers, x must equal exactly 8 and y must equal exactly 9. Thus $\frac{17}{y} = \frac{17}{9} = 1\frac{8}{9} = 1.88\bar{8}$. Therefore the correct answer is 8.

3. **30**: Assigning a variable (let's use x) to the desired quantity, the question can be translated as follows: $\frac{x}{4} = \frac{3}{5} \times 12\frac{1}{2}$. What is x?

$$\frac{x}{4} = \frac{3}{5} \times 12\frac{1}{2}$$

$$\frac{x}{4} = \frac{3}{5} \times \frac{25}{2}$$

$$\frac{x}{4} = \frac{75}{10}$$

$$x = \frac{75}{10} \times 4 = \frac{300}{10} = 30.$$

4. **75**: We should assign a variable (lets use x) to represent the unknown quantity in this percents word problem. The question can then be translated as follows:

$$\frac{x}{100} \times 6 = (15\%) \times (150\%) \times 20$$

$$\frac{x}{100} \times 6 = 0.15 \times 1.5 \times 20$$

Use your calculator to simplify:

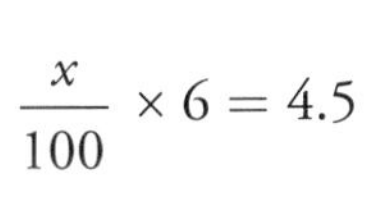

$$\frac{x}{100} \times 6 = 4.5$$

$$\frac{x}{100} = 0.75$$

$$x = 75$$

Note that the phrase "what percent" is expressed here as a variable (x) over 100. We could simply use x without a fraction, but we would need to remember to convert the value to a percentage point value at the end.

$x \times 6 = 0.15 \times 1.5 \times 20$
$x \times 6 = 4.5$
$x = 0.75$
$x = 75\% = 75$ percentage points

5. **D:** The information given in the question can be represented as follows. First, let's determine the total number of professors at the symposium, assigning a variable (in this case, x) to represent the unknown quantity:

$(20\%) \times x + (60\%) \times x + 12 = x$
$0.2x + 0.6x + 12 = x$
$0.8x + 12 = x$
$12 = 0.2x$
$60 = x$

There are 60 professors at the symposium. If 20 of them wear glasses, then 40 do not. $\frac{40}{60} = \frac{2}{3}$, or $66\frac{2}{3}\%$

6. **C:** $5^3 10^2 = 125 \times 100 = 12{,}\underline{5}00$, which has a hundreds digit of 5. $\frac{1}{4} = 0.2\underline{5}$, which has a hundredths digit of 5. Therefore, **the two quantities are equal;** the correct answer is C.

Note that in this case, it is not strictly necessary to calculate the value of the expression on the left. All powers of 5 end in 5 (5; 25; 125; 625; 3,125; etc.), and since we are multiplying the 5 term by 100, its units digit will end up in the hundreds place of the final number (500; 2,500; 12,500; 62,500; 312,500, etc.)

7. **97 babies:**

Percent change is calculated using the following formula, where B_N is the number of babies born in the new year, and B_O is the number of babies born in the original (previous) year:

$$\left(1 + \frac{\text{Percent Change}}{100}\right)B_O = B_N$$

Therefore: 

Babies in 2006 = Babies in 2005 × (1 – 0.088) = 102 × (0.912) = 93 babies (Rounded to nearest integer)
Babies in 2007 = Babies in 2006 × (1 + 0.075) = 93 × (1.075) = 100 babies (Rounded to nearest integer)
Babies in 2008 = Babies in 2007 × (1 + 0.020) = 100 × (1.020) = 102 babies (Rounded to nearest integer)
Babies in 2009 = Babies in 2008 × (1 – 0.049) = 102 × (0.951) = 97 babies (Rounded to nearest integer)

One could also solve this using the calculator. One would input as follows: 

$102 \times 0.912 \times 1.075 \times 1.020 \times 0.951 = 97.003$. However, babies only come in whole numbers, so the correct response is 97.

Note that the original table specifies that the percentage change in births each year is rounded to the nearest tenth of a percent. Therefore, for example, the actual percent change in 2006 is not necessarily precisely −8.8%—it could range anywhere between −8.85% and −8.75%. Therefore the actual number of babies born in 2006 could range between 102 × (0.9115) = 92.97 babies and 102 × (0.9125) = 93.08 babies. Following the whole percentage chain all the way through the calculation, however, the lowest possible result for 2009 is 96.81 babies and the highest possible result is 97.20 babies. Therefore, the precise number of births in 2009 must be exactly 97.

If the percent change were reported with less accuracy (for example, being rounded to the nearest percent), a wider range of possibilities could occur: the correct answer could be somewhere between 95 and 99 babies.

8. **B**: The first fraction expression can be rewritten as $\left(\frac{5x}{7}\right)\left(\frac{2x+3}{8}\right)=\frac{10x^2+15x}{56}$ (notice that we've flipped the fraction in the denominator in going from division to multiplication).

The second fraction expression can be rewritten as $\frac{7x^2}{28}+\frac{32x}{28}=\frac{7x^2+32x}{28}$ (with 28 as the least common multiple of 4 and 7). In order to compare the two fractions, it is convenient to set the denominators equal to each other. This can be achieved by multiplying the numerator and denominator of $\frac{7x^2+32x}{28}$ by 2, giving us $\frac{14x^2+64x}{56}$. We can now easily compare the numerators of the two expressions.

Since we are told that x is a positive integer, we can conclude that $14x^2 + 64x > 10x^2 + 15x$, because $14x^2$ will always be greater than $10x^2$, and $64x$ will always be larger than $15x$, whenever x is positive. Therefore **Quantity B is greater.**

9. $\mathbf{\frac{7}{18}}$ **(or any mathematical equivalent):** First, transform the mixed numbers into improper fractions:

$$\frac{3\frac{1}{4}-2\frac{2}{3}}{1\frac{1}{2}}=\frac{\frac{13}{4}-\frac{8}{3}}{\frac{3}{2}}$$

At this point, we can perform the subtraction in the numerator by expressing both fractions in terms of a common denominator. We can then divide the result by the fraction in the denominator. However, an even faster way to arrive at the answer is to determine the common denominator of *all* fractions in the calculation (including those in the denominator), and then multiply both the numerator and the denominator of the main fraction by that number. This clears the top and bottom fractions at once and leaves us with just integers. In this case, the common denominator of all fractions is 12. Therefore:

$$\frac{\frac{13}{4}-\frac{8}{3}}{\frac{3}{2}}\times\frac{12}{12}=\frac{13\times3-8\times4}{3\times6}=\frac{39-32}{18}=\frac{7}{18}$$

Note that any equivalent fraction, such as $\frac{14}{36}$ or $\frac{70}{180}$, is also an acceptable answer.

10. **C**: To perform the calculation in Quantity A as easily as possible, first find the least common denominator of all the fractions in the expression. In this case, the least common denominator is 6. Then, multiply the numerator and the denominator by that common denominator to eliminate all the fractions at once.

$$\frac{\frac{3}{2}+\frac{5}{6}}{\frac{4}{2}+\frac{4}{6}}\times\frac{6}{6}=\frac{3\times3+5}{4\times3+4}=\frac{14}{16}=\mathbf{\frac{7}{8}} \qquad \mathbf{\frac{7}{8}}$$

Therefore **the two quantities are equal.**

11. **C:** To find the ratio of two quantities, we divide the first by the second. In this case, the second quantity is a fraction, so we invert it and multiply. Also note that "of" means multiplication.

$$\frac{\frac{2}{5}\times4}{\frac{4}{3}\times5}=\frac{\frac{8}{5}}{\frac{20}{3}}=\frac{8}{5}\times\frac{3}{20}$$

At this point we could just multiply the numerators and denominators separately and then simplify the resulting fraction. However, it is preferable to cancel common factors before the multiplication, so as to minimize the amount of work we need to do to solve the problem. In this case, the 8 in the numerator and the 20 in the denominator both have a factor of 4, so we can cancel those out to obtain:

$$\frac{8}{5}\times\frac{3}{20}=\frac{\cancel{2}\times\cancel{2}\times2}{5}\times\frac{3}{\cancel{2}\times\cancel{2}\times5}=\frac{2}{5}\times\frac{3}{5}=\frac{6}{25}$$

12. $\mathbf{\frac{33}{70}}$ **(or any mathematical equivalent):** In order to calculate the fraction of all workers that are over 35, we can reason as follows: $\frac{3}{5}$ of the workers are men, and of those, $\frac{1}{2}$ are over 35. Thus $\frac{3}{5}\times\frac{1}{2}=\frac{3}{10}$ of the workers are men over 35. Similarly, $\left(1-\frac{3}{5}\right)\times\frac{3}{7}=\frac{6}{35}$ of the workers are women over 35. The total fraction of workers over 35, noting that the least common denominator of 10 and 35 is 70, is given by:

$$\frac{3}{10}+\frac{6}{35}=\frac{3\times7}{70}+\frac{6\times2}{70}=\frac{21+12}{70}=\frac{33}{70}$$

Another approach is to use a Smart Number. Given the various fractions in the problem, a number that is divisible by all the denominators is 70. Thus, we might assume that there are a total of 70 workers in the department, and solve for how many of the 70 are men and how many are women. We can then determine how many of those are over 35 years of age. The obvious advantage of the Smart Number approach is that it avoids fraction addition altogether.

Note also that any equivalent fraction, such as $\frac{66}{140}$ or $\frac{330}{700}$, is also an acceptable answer.

13. **A**: We can apply multipliers to q to account for the successive percent changes it undergoes. The multiplier for a 1% increase is $1+\frac{1}{100}=1.01$. Thus, raising q by 1% turns it into $1.01q$. This quantity must then be multiplied by 0.99, which is the multiplier for a 1% decrease. The end result for Quantity A is thus $(0.99) \times (1.01) \times q = 0.9999q$. We can perform similar calculations in Quantity B, where q is lowered and then raised by 2%:

$(0.99) \times (1.01) \times q = \mathbf{0.9999}\boldsymbol{q}$ $(0.98) \times (1.02) \times q = \mathbf{0.9996}\boldsymbol{q}$

Thus **Quantity A is greater**, because q is a positive number.

An alternative to this approach is to use the Smart Number of 100 for the initial value of q. In that case, the quantities are found to be:

99.99 **99.96**

14. **C:** The two most populous counties in State X are A and B, which together account for 31% + 22% = 53% of the total population. This means that all other counties make up 100% – 53% = 47% of the total population. It is tempting to answer that the difference is 53% – 47% = 6%, but that would be incorrect. This is because 6% is the difference *relative to the total population of the state*. The problem asks us to express the answer relative to the *combined population of all other counties*. The correct calculation is therefore:

$$\% \text{ greater} = \frac{53\% - 47\%}{47\%} \times 100 \approx 13.$$

15. **D:** First we should write x as a decimal in order to calculate x^2. This gives $x = 0.8$ and $x^2 = 0.64$. The percent difference is equal to the actual difference divided by the basis of the comparison. The basis of the comparison is what follows the word "than" in the problem statement:

$$\% \text{ difference} = \frac{0.8 - 0.64}{0.64} = \frac{0.16}{0.64} = \frac{1}{4} = 25\%$$

Note that, if the problem had instead been worded as "by what percent is 0.64 smaller than 0.8?", the basis of the comparison would have been 0.8. The answer in that case would have been 20%.

16. **A**: Using the acronym PEMDAS to remember the proper order of operations, we first perform the operations inside the parentheses:

$$\frac{\frac{3}{8} \div \left(\frac{3}{4} - \frac{1}{3}\right)}{\frac{2}{5}} = \frac{\frac{3}{8} \div \left(\frac{9}{12} - \frac{4}{12}\right)}{\frac{2}{5}} = \frac{\frac{3}{8} \div \frac{5}{12}}{\frac{2}{5}}$$

All that remains is division. To divide fractions, we "flip" them, then multiply:

$$\frac{\frac{3}{8} \div \frac{5}{12}}{\frac{2}{5}} = \frac{3}{8} \times \frac{12}{5} \times \frac{5}{2}$$

Finally, we can cancel/reduce terms to simplify the math when multiplying through:

$$\frac{3}{8} \times \frac{\overset{6}{\cancel{12}}}{\underset{1}{\cancel{5}}} \times \frac{\overset{1}{\cancel{5}}}{\underset{1}{\cancel{2}}} = \frac{3}{\underset{4}{\cancel{8}}} \times \frac{\overset{3}{\cancel{6}}}{1} = \frac{9}{4} = 2.25.$$

17. **B:** Let us define $x = 0.\overline{12} = 0.121212...$ Then, we can see that $100x = 12.121212... = 12 + x$. Solving for x, we get $100x - x = 12.\overline{12} - 0.\overline{12}$, so $99x = 12$. Thus $x = \frac{12}{99} = \frac{4}{33}$.

Alternatively, you could calculate the fraction in each Choice in the Calculator on the screen and see that only Choice B produces the desired non–terminating decimal pattern.

Note that the pattern demonstrated above for non–terminating decimals can be generalized: a repeating decimal of the form $0.\overline{a}$ will equal $\frac{a}{9}$; a repeating decimal of the form $0.\overline{ab}$ will equal $\frac{ab}{99}$; a repeating decimal of the form $0.\overline{abc}$ will equal $\frac{abc}{999}$, and so on. Note that in this representation a, b, and c are digits of a decimal, so that (for instance) ab does *not* stand for the product of a and b.

18. **C:** The key to solving this problem successfully is to multiply the terms that are grouped next to each other in parentheses *first*, so that you do not make an error:

$$\frac{(1)(0.1)+(2)(0.2)+(3)(0.3)+(4)(0.4)}{(1)(0.1)+\left(\frac{1}{2}\right)(0.2)+\left(\frac{1}{3}\right)(0.3)+\left(\frac{1}{4}\right)(0.4)} = \frac{0.1+0.4+0.9+1.6}{0.1+0.1+0.1+0.1} = \frac{3}{0.4} = \frac{30}{4} = 7.5.$$

19. **B:** First, let us solve for the numerator:

$$\frac{5}{3} - \frac{1}{6} = \frac{10-1}{6} = \frac{9}{6} = \frac{3}{2}.$$

Next, let us solve for the denominator:

$$\frac{0.3^2}{0.02} = \frac{0.09}{0.02} = \frac{9}{2}.$$

Therefore, $\dfrac{\frac{5}{3} - \frac{1}{6}}{\frac{0.3^2}{0.2}} = \dfrac{\frac{3}{2}}{\frac{9}{2}} = \dfrac{1}{3}.$

20. **C, D, E, and F**: For all packages weighing above 5 pounds, the shipping cost C is given by $C = \$2.80 + \$0.25 \times (P - 5)$, where P is the weight, in pounds, of the package. According to the problem, the cost to ship Package A is greater than \$7 and less than \$8. Therefore:

$$\$7 < \$2.80 + \$0.25 \times (P - 5) < \$8$$

We can subtract \$2.80 from all 3 expressions in the compound inequality:

$$\$4.20 < \$0.25 \times (P - 5) < \$5.20$$

Then, we can divide all 3 expressions by \$0.25:

$$\frac{\$4.20}{\$0.25} < P - 5 < \frac{\$5.20}{\$0.25}$$

$$16.8 < P - 5 < 20.8$$

Finally, we can add 5 to all 3 expressions:

$$21.8 < P < 25.8$$

The package can thus weigh 22, 23, 24, or 25 pounds.

FDPs: Hard Practice Question Solutions

1. **B**: In order to evaluate the fraction in Quantity A, we should simplify as follows:

$$\left[\frac{\frac{(x^2-9)}{3}}{\frac{x+3}{8}}\right]^{-1}=\frac{\frac{x+3}{8}}{\frac{(x^2-9)}{3}}=\frac{\frac{x+3}{8}}{\frac{(x+3)(x-3)}{3}}=\frac{3\cancel{(x+3)}}{8\cancel{(x+3)}(x-3)}=\frac{3}{8(x-3)}$$

Since $x > 4$, the denominator in Quantity A must be greater than 8; since the numerators in Quantity A and Quantity B are the same and the denominator in Quantity A is larger, **Quantity B must be greater.**

Note the use of the difference of squares special product to factor the $(x^2 - 9)$ expression in Quantity A. Also note that in simplifying the fraction we flipped the fraction $\frac{(x+3)(x-3)}{3}$ as we went from division to multiplication. Finally, notice that any number raised to the power of -1 is equal to its reciprocal.

2. **B:** It is fairly easy to see that the value of y exactly doubled between years 1 and 2, implying a percent increase of 100%. In order to calculate the percentage increase more formally, we use the following formula:

Percent change in $y = \frac{y_2 - y_1}{y_1} = \frac{400-200}{200} = 1 = 100\%$.

At this point we know that the percent increase in y must be greater than the percent decrease in x, because a 100% decrease in x would imply that $x_2 = 0$, and any decrease greater than 100% would result in x_2 being negative. Neither of these is true, so **Quantity B is greater.** Still, it is a worthwhile exercise to actually compute the value of the percent decrease in x:

Percent change in $x =$

$$\frac{x_2-x_1}{x_1}=\frac{2^5\times5^5-2^1\times10^5}{2^1\times10^5}=\frac{2^5\times5^5-2^1\times(2^5\times5^5)}{2^1\times(2^5\times5^5)}=\frac{(2^5\times5^5)\times(1-2^1)}{2^1\times(2^5\times5^5)}=\frac{1-2}{2}=-\frac{1}{2}=-50\%.$$

This is a 50% decrease (notice that the negative percent change implies a decrease, but we are trying to measure the positive decrease, so Quantity A is actually 50%, not -50%).

3. **C**: To evaluate the two quantities, it is not necessary to determine a value for the exponential expressions above, or even for x. 6 raised to any integer value will produce a units digit of 6, as the units digit will always be determined by the product $6 \times 6 = 3\underline{6}$. You can see this pattern by trying the first few exponents of 6 (or any other number ending in 6) on your onscreen Calculator. Simply continue to multiply each result by 6. The first few results are as follows:

$6^1 = \underline{6}$
$6^2 = 3\underline{6}$
$6^3 = 21\underline{6}$
$6^4 = 1{,}29\underline{6}$

4^{2x} will always produce a units digit of 6, as well. The units digits of successive powers of 4 follow a two–step pattern, as we can see by testing the first few exponents of 4 (or any integer with a ones digit of 4) on the Calculator:

$4^1 = 4$
$4^2 = 16$
$4^3 = 64$
$4^4 = 256$

Note that odd exponents of 4 always end in 4, and even exponents of 4 always end in 6. Since $2x$ must be even, 4^{2x} will have a units digit of 6. **The two quantities are equal.**

4. **A, B and C:** The GRE entices test takers to make unwarranted assumptions. Here, it is important that we *not assume* that x is an integer. If it were, then Choice A is the only possibility. For example, $\sqrt{4} > \frac{\sqrt{4}}{4}$ and no integer would result in Choices B or C being correct. However, x could be a fraction and then the other choices (Choices B and C) are possible. For example, $\sqrt{\frac{1}{4}} > \frac{\sqrt{\frac{1}{4}}}{\frac{1}{4}}$, because $\sqrt{\frac{1}{4}} = \frac{1}{2}$ and $\frac{\sqrt{\frac{1}{4}}}{\frac{1}{4}} = \frac{\frac{1}{2}}{\frac{1}{4}} = 2$. Furthermore, $\sqrt{\frac{1}{4}} > \frac{1}{4}$. Thus both Choices B and C are possible.

Picking numbers such as the above is the best approach, but again, the key is not to assume beyond the exact words given in the question stem.

5. **C, D, E,** and **F**: The information in the question can be translated as follows:

$$\frac{150}{100} \times (100) \leq x \leq \frac{200}{100} \times (100)$$
$$0.2x \leq y \leq 0.5x$$

Note that in the first inequality, we are looking at a percent *increase*, which is equal to x plus the decimal equivalent of the percent change. In the second inequality, y is a percentage of x, so we multiply x by the given percentages.

From the first inequality, we can determine that x is between 150 and 200. Now we just need to find the possible range for y. It can be anywhere from 20% to 50% of *any number in the range* for x. The smallest possible value for y is 20% of the smallest number (150), and the largest possible value is 50% of the largest number (200). Thus $0.2 \times 150 \leq y \leq 0.5 \times 200$, or $30 \leq y \leq 100$. Of the numbers listed, 40, 60, 90, and 100 fall within this range.

6. **25**: In order to avoid selling at a loss, the shopkeeper must increase the selling price of the birdfeeders to match their cost. To determine the percent increase required, we must first find the required selling price by calculating the new cost of the birdfeeders. Then we must express this dollar increase as a percentage of the original selling price of \$18 by applying the following formula, where P_N and P_O represent the new and original prices, respectively:

$$\text{Percent change in price} = \frac{P_N - P_O}{P_O} \times 100.$$

The cost of the birdfeeders (\$10) increased by 50% twice in a row. We can express this as \$10 × (1 + 50%) × (1 + 50%) = \$10 × 1.5 × 1.5, which equals \$22.50. (Note that two successive 50% increases is *not* the same as a single increase of 100%!)

Now we can plug the new price into our percent increase formula:

Percent change in price $= \frac{P_N - P_O}{P_O} \times 100 = \frac{\$22.50 - \$18}{\$18} \times 100 = \frac{4.5}{18} \times 100 = 25$

7. **A:** For every four consecutive integers, exactly one must be divisible by 4. Since neither x nor $x - 1$ is divisible by 4, then either $x + 1$ or $x + 2$ must be divisible by 4 (but not both).

Suppose that $x + 1$ is divisible by 4, i.e., $\frac{x+1}{4} = Q + \frac{0}{4}$, where Q is an integer and zero is the remainder. This implies that $\frac{x}{4} + \frac{1}{4} = Q$, which implies that $\frac{x}{4} = Z + \frac{3}{4}$ or $Z.75$ (where $Z = Q - 1$) and that the tenths digit (7) is greater than the hundredths digit (5).

Suppose instead that $x + 2$ is divisible by 4. Using the same logic we can conclude that $\frac{x}{4} = Z.50$ (again, where $Z = Q - 1$) and once again the tenths digit (5) is greater than the hundredths digit (0).

Since either $x + 1$ or $x + 2$ must be divisible by 4, the tenths digit of $\frac{x}{4}$ must be greater than the hundredths digit.

8. **3:** Last digit (or units digit) problems should be approached in a standard way, which is to avoid lengthy calculation and instead look for patterns in the last digit as a number is raised to successive powers. Even the raising to successive powers can be streamlined by limiting the calculation to the last digit and ignoring the rest of the digits. For powers of 3, the process goes as follows:

k	**Last digit of 3^k**
1	3
2	9
3	7 (because 9 × 3 = 2<u>7</u>)
4	1 (because 7 × 3 = 2<u>1</u>)
5	3

Note that the last digit of 3^k will repeat from this point onward; it will continually cycle through 3–9–7–1, in that order. That is, every power of 3 that has a multiple of 4 as the exponent will have its last digit equal to 1. This means that 3^{24} will end in a 1. Therefore 3^{23} will end in a 7.

We can proceed similarly for powers of 2:

k	**Last digit of 2^k**
1	2
2	4
3	8
4	6 (because 8 × 2 =1<u>6</u>)
5	2 (because 6 × 2 =1<u>2</u>)

Again, we note that there is a cycle: every power of 2 that has a multiple of 4 as the exponent will have its last digit equal to 6. This means that 2^{20} will end in 6, 2^{19} will end in an 8, and 2^{18} will end in 4. Finally, then, $3^{23} - 2^{18}$ will end in $7 - 4 = 3$.

9. B, C, and **D:** In order to answer this question, we need to determine the possible extremes in the percent salary increase. The least possible increase would be from the highest 1990 salary to the lowest 2000 salary:

$$\text{Minimum percent increase} = \frac{\$70{,}000 - \$55{,}000}{\$55{,}000} = \frac{\$15{,}000}{\$55{,}000} = \frac{3}{11} \approx 27.3\%$$

The greatest possible increase, meanwhile, would be from the lowest 1990 salary to the highest 2000 salary:

$$\text{Maximum percent increase} = \frac{\$110{,}000 - \$40{,}000}{\$40{,}000} = \frac{\$70{,}000}{\$40{,}000} = \frac{7}{4} = 175\%$$

Percent increases that fall between these two extremes are possible.

10. **C:** For the first two statements we need only to work with fractions. Choice A will be true if $\frac{3}{8}$ is greater than $\frac{2}{5}$. We can quickly compare the two by cross–multiplying the numerators and denominators, and writing the results next to the numerators. Whichever product is greater will indicate which fraction is greater:

$\frac{3}{8}$ versus $\frac{2}{5}$ is the same as 3×5 versus 2×8: 15 is less than 16, so $\frac{3}{8}$ is less than $\frac{2}{5}$. Choice A is therefore incorrect.

As for Choice B, the fraction of students who took French, Geography or both can be determined by adding the individual fractions for each and then *subtracting* the fraction that represents the overlap (to eliminate the double–counting of the students who are in both classes). This gives $\frac{3}{8} + \frac{2}{5} - \frac{1}{4} = \frac{3 \times 5 + 2 \times 8 - 1 \times 10}{40} = \frac{21}{40}$ as the fraction of students who took **either class or both.** Therefore, the fraction of students who took *neither* French *nor* Geography must equal $1 - \frac{21}{40} = \frac{19}{40}$. This is greater than $\frac{1}{4}$, which is the fraction of students who took both. Thus Choice B is also incorrect.

Finally, to check Choice C, we first need to compute the fraction of students who took only Geography and then multiply by 420 to determine the actual number of such students. The fraction of students who took Geography but not French is found by subtracting the overlap from the fraction of students who took Geography: $\frac{2}{5} - \frac{1}{4} = \frac{2 \times 4 - 1 \times 5}{20} = \frac{3}{20}$. The number of such students is given by $420 \times \frac{3}{20} = 3 \times \frac{420}{20} = 3 \times 21 = 63$.

Therefore Choice C is correct.

11. **E:** We can make a table to summarize what we know about the "Before" and "After" states. Even though the problem asks for Jennifer's number of stamps, it actually is more convenient to let Peter's number of stamps be our unknown, because we are told how Jennifer's number of stamps relates to Peter's: "Jennifer has 40% more than Peter." Let P denote the number of stamps that Peter has before the transfer of 45 stamps. We then have:

	Before	**After**
Peter	P	$P + 45$
Jennifer	$1.4P$	$1.4P - 45$

Note in the table that 40% more than P translates to: $P+\frac{40}{100}P=P+0.4P=1.4P$. The "After" column is filled out by accounting for the transfer of 45 stamps.

At this point, we can invoke the second given fact, namely that Peter will have 10% more stamps than Jennifer in the "after" state. Just as "40% more than" translated to a coefficient of 1.4, "10% more than" will translate to a coefficient of 1.1. Thus we can equate Peter's "After" total with 1.1 × Jennifer's "After" total:

$$P + 45 = 1.1(1.4P - 45) = 1.54P - 49.5$$

Collecting terms yields:

$$94.5 = 1.54P - P = 0.54P$$

This is a calculation best done using the Calculator: $P=\frac{94.5}{0.54}=\frac{9,450}{54}=\frac{1,050}{6}=175.$ We must remember, however, that the question asked us for how many stamps *Jennifer* has at the beginning. This is also best computed with the Calculator: $1.4 \times 175 = 245$.

12. **17:** Assume that the list price of the boat is $100. The dealer buys the boat for $90. He then sells it for $\$90+\frac{30}{100}\times\$90=1.3\times\$90=\$117$. From this, we can see that the dealer would have made a profit of $17 if he had bought the boat at list price. His profit would then have been 17% over the list price, which would be entered as the whole number 17. (Note that choosing $100 as the list price makes the calculation of the profit very convenient—a profit of $17 is the same as a profit of 17%.)

13. **B:** The problem statement can be translated verbatim. "Percent" means divide by 100, "of" means multiply, and "P% less than P" means subtract P% of P from P:

$$\frac{P}{100}(3P)=P-\frac{P}{100}\times P=\left(1-\frac{P}{100}\right)P$$

The equation can be simplified by multiplying both sides by 100. This eliminates the fractions and results in:

$$3P^2=(100-P)P=100P-P^2$$

Collecting the squared terms on the left and dividing by P (which is not zero), we obtain:

$$4P^2 = 100P$$
$$4P = 100$$
$$P = 25$$

14. **A:** The compound interest formula can be used to calculate the future value F resulting from an initial principal of P plus interest accruing at a rate of r, compounded quarterly for n years. This formula is given by:

$$F = P\left(1+\frac{r}{4}\right)^{4n}$$

Where r is to be expressed as the decimal equivalent of the interest rate. (Note the use of 4 in the formula, to represent the quarterly compounding, i.e., 4 times per year.)

In this problem, $r = 0.08$ and $n = 3$, so \$12,000 $= P \times (1.02)^{12}$, or $P = \frac{\$12{,}000}{1.02^{12}}$.

Using the Calculator, we can obtain 1.02^{12} most efficiently by using the shortcut "× =" for squaring a number. We first square 1.02 and then multiply it by 1.02 again to get 1.02^3. We then square 1.02^3 to get 1.02^6, then square that number to get 1.02^{12}: 1.02×=×1.02 =×=×=yields≈1.26824. Making sure to clear the memory first by pressing *MC* if needed, we store this number in the Calculator's memory via *M+*, and finally divide \$12,000 by the number in memory to arrive at an approximate answer of $P =$ \$9,462.

15. **\$9,274:** The compound interest formula yields the final value of principal P plus interest at a rate of r, compounded annually for n years. The formula is: $F = P(1 + r)^n$, where r is expressed as the decimal equivalent of the annual percentage interest rate. In this case, $r = 0.06$.

The first year's investment will compound 4 times, and result in a final value of \$2,000 × 1.06^4.
The second year's investment will compound 3 times, and result in a final value of \$2,000 × 1.06^3.
The third year's investment will compound twice, and result in a final value of \$2,000 × 1.06^2.
The fourth year's investment will compound once, and result in a final value of \$2,000 × 1.06.

Thus, the total accumulation at the end of the fourth year, therefore, will equal:

$$\$2{,}000 \times (1.06^4 + 1.06^3 + 1.06^2 + 1.06) \approx \$2{,}000 \times 4.63709 \approx \$9{,}274$$

16. **C:** First, we can group the exponents in the denominators to factor out the common terms:

$$\frac{1}{2^{11}5^7}+\frac{1}{2^7 5^{11}}=\frac{1}{2^4}\cdot\frac{1}{2^7 5^7}+\frac{1}{5^4}\cdot\frac{1}{2^7 5^7}=\left(\frac{1}{2^4}+\frac{1}{5^4}\right)\cdot\frac{1}{2^7 5^7}$$

Now, $\frac{1}{2^7 5^7}=\frac{1}{10^7}=10^{-7}$ will have 6 zeros after the decimal point prior to the first non–zero digit. We next need to determine how the leading term $\frac{1}{2^4}+\frac{1}{5^4}$ affects the result:

$$\frac{1}{2^4}+\frac{1}{5^4}=\frac{5^4+2^4}{2^4 5^4}=\frac{625+16}{10^4}=\frac{641}{10{,}000}=0.0641$$

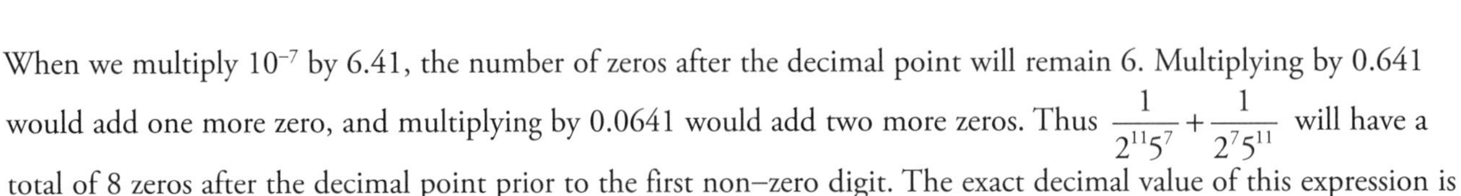

When we multiply 10^{-7} by 6.41, the number of zeros after the decimal point will remain 6. Multiplying by 0.641 would add one more zero, and multiplying by 0.0641 would add two more zeros. Thus $\frac{1}{2^{11}5^7}+\frac{1}{2^7 5^{11}}$ will have a total of 8 zeros after the decimal point prior to the first non–zero digit. The exact decimal value of this expression is 0.00000000641.

17. **E:** 10% less than 2 means (1 – 10%) = 90% of 2. (It does not mean 2 – 0.1.) Thus, we are looking for the expression whose value is 0.9 × 2 = 1.8. We can do the calculations using simple decimal representations of the various fractions. In some cases these will be exact; in other cases they will be approximate. The advantage of this approach is that it avoids the effort involved in doing fraction arithmetic via common denominators. We get:

(A) $\frac{1}{2}+1\frac{2}{5}=0.5+1.4=1.9$ (Exact, incorrect)

(B) $\frac{5}{6}+1\frac{1}{3}\approx 0.8+1.3=2.1$ (Approximate but an underestimate, therefore incorrect)

(C) $1\frac{2}{3}+\frac{3}{10}\approx 1.6+0.3=1.9$ (Approximate but an underestimate, therefore incorrect)

(D) $2\frac{1}{6}-\frac{2}{5}\approx 2.2-0.4=1.8$ (Approximate but an overestimate, therefore incorrect)

(E) $2\frac{1}{2}-\frac{7}{10}=2.5-0.7=1.8$ (Exact, correct)

18. **B, C, and D:** The largest allowable area of the rectangle is given by:

$$\left(3+\frac{1}{2}\right)\left(2+\frac{1}{4}\right)=\left(\frac{7}{2}\right)\left(\frac{9}{4}\right)=\frac{63}{8} \text{ or } 7\frac{7}{8} \text{ square inches}$$

The percent difference between this area and 6 square inches is equal to $\frac{7\frac{7}{8}-6}{6}=31.25\%$. Thus, Choices C and D are possible.

Meanwhile, the smallest allowable area of the rectangle is given by:

$$\left(3-\frac{1}{2}\right)\left(2-\frac{1}{4}\right)=\left(\frac{5}{2}\right)\left(\frac{7}{4}\right)=\frac{35}{8} \text{ or } 4\frac{3}{8} \text{ square inches}$$

The percent difference between this area and 6 square inches is equal to $\frac{4\frac{3}{8}-6}{6}\approx -27.1\%$. Thus, Choice B is possible, while Choice A is not.

19. **A:** In order to determine the value of *B*, the ones (units) digit of the product, we need only look at the product of the units digits of 12,34**5** and 6,78**9**: 5 × 9 = 4**5**, so *B* = **5**.

To determine the value of *A*, the tens digit, we need to look at the product of the tens *and* ones digits of 12,345 × 6,789: 45 × 89 = 2,205, so *A* = 0.

Therefore *AB* = 0 × 5 = 0.

20. C, E, G, and I: Whenever a question asks for the value of the ones (units) digit result of multiplication, one only needs to calculate the product of the units digits of the original operands. For example, $49 \times 143 = 7{,}007$, but one can find the units digit of the product by taking $9 \times 3 = 2\underline{7}$.

Thus to find the possible units digits of 98^x, one need only observe the units digits of successive powers of 8^x:

x	**Units digit of 8*x***
1	8
2	4 ($8 \times 8 = 6\underline{4}$)
3	2 ($4 \times 8 = 3\underline{2}$)
4	6 ($2 \times 8 = 1\underline{6}$)
5	8 ($6 \times 8 = 4\underline{8}$)

Thus the pattern of 8, 2, 4, 6 will repeat every 4 successive powers, and the units digit of 98^x can only ever be 2, 4, 6, or 8.

Notice also that we can use Primes & Divisibility and Odd/Even concepts to eliminate all but the correct Choices. All odd digits can be eliminated (since 98 is an even number, 98 to any power will be an even number and thus end in an even digit), as can 0 (any integer ending in 0 is divisible by 10, which means that the integer has both a 2 and a 5 in its prime factorization; the prime factorization of 98 is $2 \times 7 \times 7$, so 98 raised to any power will *never* have a 5 in its prime factorization and thus cannot be divisible by 10).